P9-CMY-316

Cultivating *Delight*

ALSO BY DIANE ACKERMAN

Deep Play
I Praise My Destroyer
A Slender Thread
The Rarest of the Rare
A Natural History of Love
A Natural History of the Senses
The Moon by Whale Light
Jaguar of Sweet Laughter
Reverse Thunder
On Extended Wings
Lady Faustus
Twilight of the Tenderfoot
Wife of Light
The Planets: A Cosmic Pastoral

FOR CHILDREN

Monk Seal Hideaway
Bats: Shadows in the Night

ANTHOLOGY

The Book of Love
(with Jeanne Mackin)

Cultivating *Delight*

A Natural History *of* My Garden

DIANE ACKERMAN

HarperCollins*Publishers*

HarperCollins books may be purchased for educational, business, or sales promotional use. For information, please write: Special Markets Department, HarperCollins Publishers Inc., 10 East 53rd Street, New York, NY 10022.

An extension of this copyright page may be found on page 260.

FIRST EDITION

Designed by Elliott Beard

Printed on acid-free paper

Library of Congress Cataloging-in-Publication Data

Ackerman, Diane.
Cultivating delight: A natural history of my garden / Diane Ackerman
p. cm.
ISBN 0-06-019986-5
1. Natural history. 2. Gardens. I. Title.
QH45.2.A34 2001
508—dc21 2001016607

01 02 03 04 05 ❖/RRD 10 9 8 7 6 5 4 3 2 1

The crocuses and the larch turning green every year a week before the others and the pastures red with uneaten sheep's placentas and the long summer days and the newmown hay and the wood pigeon in the morning and the cuckoo in the afternoon and the corncrake in the evening and the wasps in the jam and the smell of the gorse and the look of the gorse and the apples falling and the children walking in the dead leaves and the larch turning brown a week before the others and the chestnuts falling and the howling winds and the sea breaking over the pier and the first fires and the hooves on the road and the consumptive postman whistling "The Roses Are Blooming in Picardy" and the standard oil-lamp and of course the snow and to be sure the sleet and bless your heart the slush and every fourth year the February debacle and the endless April showers and the crocuses and then the whole bloody business starting all over again.

—SAMUEL BECKETT, *Watt*

Cultivating *Delight*

1

I plan my garden as I wish I could plan my life, with islands of surprise, color, and scent. A seductive aspect of gardening is how many rituals it requires. Uncovering the garden in the spring, for example. Replacing a broken-down metal gate with a burly wooden one. Transplanting rhododendrons to a sunnier spot. Moving the holly bushes to the side of the garage, to hide them from the deer, who nonetheless find and eat them, prickles and all. (It may be like our affection for strong peppermints, hot mustards, spicy peppers—maybe the prickles add a certain *frisson* to the deer's leafy diet.) By definition, the garden's errands can never be finished, and its time-keeping reminds us of an order older and one more complete than our own. For the worldwide regiment of gardeners, reveille sounds in spring, and from then on it's full parade march, pomp and circumstance, and ritualized tending until winter. But even then there's much to admire and learn about in the garden—the hieroglyphics of animal tracks in the snow, for instance, or the graceful arc of rose canes—and there are many strategies to plan.

Surely there is a new way to outwit the marauding deer and Japanese beetles? Gardeners understand the word *pestilence* as only medieval burghers did. Gardeners can be cultured and refined. They can be earthy, big-hearted folk who love to get their hands dirty as they dig in the sunshine. They may obsess about tidy worlds of miniature, perfectly blooming trees. They may develop a passion for jungle gardens reminiscent of Amazonia. They may specialize in making deserts bloom. They may adore the weedy mayhem of huge banks of wildflowers. They may create interflowing worlds of color and greenery, in which small meadows give way to a trellised rosebed; a moon garden with blossoms all silver or white; a water garden complete with small bridge and waterfall; a butterfly garden also visited by hummingbirds; a "flamboyant" garden filled only with red, yellow, and orange flowers; a hedge of pampas grasses whose tall plumes sway like metronomes.

Gardeners have unique preferences, which tend to reflect dramas in their personal lives, but they all share a love of natural beauty and a passion to create order, however briefly, from chaos. The garden becomes a frame for their vision of life. Whether organic or high-tech, they share a dark secret, as well. Despite their sensitivity to beauty and respect for nature, they all resort to murder and mayhem with steel-willed cunning.

Nurturing, decisive, interfering, cajoling, gardeners are eternal optimists who trust the ways of nature and believe passionately in the idea of *improvement*. As the gnarled, twisted branches of apple trees have taught them, beauty can spring in the most unlikely places. Patience, hard work, and a clever plan usually lead to success: private worlds of color, scent, and astonishing beauty. Small wonder a gardener plans her garden as she wishes she could plan her life.

Spring

Every blade of grass has its Angel
that bends over it and whispers,
"Grow, grow."

— The Talmud

2

One day, when the last snows have melted, the air tastes tinny and sweet for the first time in many months. That subtle tincture of new buds, sap, and loam I've learned to recognize as the first whiff of springtime. Suddenly a brown shape moves in the woods, then blasts into sight as it clears the fence at the bottom of the yard. A beautiful doe with russet flanks and nimble legs, she looks straight at me as I watch from the living room window, then she drops her gaze.

Like fireworks, five more deer make equally spectacular leaps, and land squarely on the lifeless grass. But once they touch earth all their buoyancy seems to vanish, and they lumber around the yard, droopy, gaunt, exhausted. A big doe lifts a hind foot to scratch her shaggy cheek. I think that's the doe I named Triangle last year because of a geometrical pattern in her coat. But at the moment, the deer are in molt, which must feel itchy, and since they don't shed evenly, their usually sleek coats look crisscrossed by small weather systems. Mainly the deer seem frantic with hunger. They start eating the dried-up lavender leaves, whose pungent smell usually keeps them at bay. They've lost their winter fat, but there's nothing in bloom. Their living larder won't be full for weeks. Desperate, they devour the bittersweet and pull bark from aspens and other trees they don't prefer. In summer's banquet, they can afford to be sloppy and eat whatever they fancy. Now, searching for the highest-protein foods, they hunt carefully among the garden's leftovers.

I love watching the deer, which always arrive like magic or miracle or the answer to an unasked question. Can there be a benediction of deer on a chilly spring morning? I think so. Their otherworldliness stops the day in its tracks, focuses it on the hypnotic beauty of nature, and then starts the day again with a rush of wonder. There is a way of sitting quietly and beholding nature which is a form of meditation or prayer, and like those healing acts it calms the spirit.

Come summer, of course, the deer will ransack my herb garden, plunder my roses, destroy the raised beds, leaving their footprints as calling cards among the decapitated flowers. They are terrorists in the garden. That's why I've planted most of my roses in a special fenced-in garden with a solid gate. It feels a little odd, being in competition with deer, but that's what comes of cultivating so much land. If we replace their vegetation with buildings and crops, we leave the deer no choice but to raid our gardens. And humans have been known to raid deer, even cultivate them as a crop. For millennia, Chinese herbals have included deer parts, favoring antler velvet, bone marrow, spinal cord, penis, undigested milk, fetus, brains, thyroid gland, and much more as remedies. If human beings were not sitting smugly atop the food chain, what curios of our bodies would others use as medicines? A thought explored in gory detail by many science fiction books and movies. These deer are beautiful and whole, not a sum of their parts, and I'm happy to share everything but prize flowers with them. But how to protect those flowers? Friends have recommended such deer deterrents as bags of hair from a barbershop, used tampons, cougar pee, mothballs, salsa, bars of smelly soap, and barking dogs. I find that an oily soap called Hinder works reasonably well, and won't harm child or animal, but you have to spray it after every rainfall. This year I finally hit on a solution: pinwheels. I planted one beside each treasured bush or flower. Spinning randomly and sparkling in the sun, they seem alive and so the deer avoid them. I also like how festive the pinwheels look, decorating the garden with color and motion and the soft whir of their blades.

But I leave all the apples from both trees just for the deer, I let them eat their fill from the raspberry vines, and I feed them in hard weather. Sometimes in colder months I leave apples beneath the twin apple trees where deer would expect to find them. For a decade, the apple trees have helped the deer survive winter. What with the changing current of El Niño and several volcanoes hurling dust high into the atmosphere, the apple trees were sparse this year and the deer found few apples beneath the snow. Despite thick, burro-like coats, they looked thin. I suppose I am "conflicted" about the deer, as psychological folk like to say. But mainly I am grateful to have these emissaries of the wild so close at hand, and when they visit all I can manage is praise.

We've worked hard to exile ourselves from nature, yet we end up longing for what we've lost: a sense of connectedness. For many homeowners, suburbanites, and travelers, backyard animals such as deer, squirrels, birds, and raccoons become an entryway to the bustling world of nature. Studying animals is easy when they're close at hand, and there's nothing like the thrill of recognizing individual animals with unique looks and personalities: the doe with the white half-circle beneath one eye, who stands up on her rear legs to pick apples off the trees; the young buck who always does a small war dance before he leaps a fence. When that happens, we lose our "us against them" attitude and start to feel part of a kingdom of neighbors. Deep in our instincts and cells we remember living wild in nature, fitting into the seamless circle of the seasons, reading the weather and landscape, facing frights and challenges. In a real sense we are out of our element now, and it's small wonder that we relish rare visits to nature—picnics, jogs, and bike rides; journeys to parks, camp grounds, and zoos.

We may feel cozy and safe in our homes, protected from both blast and predator, but we pay the price with slack muscles, weak hearts, and glum spirits. Deprived of fresh daylight, we sink low during winter months. And yet when we search for remedies to those distresses, only the artificial springs to mind: gyms, pills, lightboxes. By retreating farther and farther from nature, we lose our sense of belonging, suffer a terrible loneliness we can't name, and end up depriving ourselves of what we need to feel healthy and whole. Children know this instinctively, and quickly learn the joys of nature. When a tree stump or marsh beckons, they dive in, wide-eyed, all hands.

I suppose what we fear is loss of control, of ourselves and of our planet; and there's no doubt, nature is chaotic, random, violent, uncontainable, no matter how hard we try to outwit it. But it's also dazzling, soothing, all-embracing, and restorative. Wonder is a bulky emotion; when it fills the heart and mind there's little room for anything else. We need the intimate truths of daylight and deer.

Deer panic so easily that the only way to be among them without frightening them is to hunker down low and positively not look. Eye contact, even glancing, may distress them. Most often, wild animals make eye contact only when they wish to fight, eat, or mate. If you seem to be ignoring them, you pose little threat. And so I steal out, bent low, carrying a sliced-up peach I place on the grass near the apple trees, then, still without making eye contact, I creep back indoors. Soon two deer sniff their way to the treasure, so unfamiliar yet so sweet, and stand eating peach slices with juice dripping from their mouths. I've noticed that squirrels seem comforted by the sound of my voice when I'm among them, but deer require silence.

One noon last summer, I saw two fawns sitting on the grass in the shade of a large tree in my front yard. Quietly I crept out with concealed purpose—I walked easily across the yard, as if on an errand unrelated to the deer. Because I seemed preoccupied by human things, they watched me, ever alert, but didn't bother to stir as I sat down in the grass near them, averting my eyes, picking a blade of grass or two, only now and then studying them with long, thick glances. A passing car startled them and they half-stood, then settled down again. Deer don't fold their legs like dogs, but slide down over the tops of their knees like camels. As I continued my mock grazing, they curled up and snoozed.

Where was their mother, and wasn't she afraid to leave them

alone among humans? Typically, a doe will have one or two fawns and hide them in a secluded spot while she forages, returning only to nurse them. During the first few weeks of life, fawns don't give off much scent. Small, quiet, camouflaged, and nearly odorless, they're not easily discovered, so the mother may drift off without much concern. On the other hand, I might well have been in her sights, and dismissed as another creature out grazing in the sun. Humans are familiars to suburban deer.

By midsummer the fawns were eating a vegetarian diet which included hundreds of species of buds and leaves, and they were roaming a wide area. They left scent clouds in the forest. Late in the day, I would often see them wander into open areas to nibble broad-leafed plants, watch them position themselves downwind, slowly roaming upwind, always alert for the smell of predators. Deer tend to twitch their tails just before lifting their heads, so whenever I see the tail-twitch, I stand still and drop my eyes. If I scare them, they'll pronk away in an awkward, tail-flashing pantomime that says, "I know you see me, but I'm too fast for you! Too strong for you! It's no use chasing me!" To my mind, prey calling attention to itself as a bluff seems dicey, but apparently gazelles do the same thing.

Now, nearly a year later, the fawns are grown, their speckles have disappeared and their bellies look gaunt. The lovely reddish-brown coats they wore as gawky juveniles have been replaced by the somber gray-brown of adulthood. These coats are more loosely woven, but each hair is hollow so that trapped air will heat up like a comforter on chilly nights. I cannot tell if they're male or female—not from a distance, anyway. Their mother is young, so most likely the fawns are female; deer tend to produce females early in their breeding life. Later on, as a doe ages, it's advantageous to have males and fuss over them, suckling them longer, making sure they're strong so that her genes will flourish as her offspring grow to best their rivals and dominate harems. Many animals are instinctively able to choose the sex of their offspring. Do humans make such unconscious choices? I wonder.

Remembering what I was saving for the squirrel and bird feeders, I grab a jacket and hurry into the garage, returning soon with six ears of corn. Then slipping slowly out back, I toss the corn across the yard. The deer stare squarely at it from a distance, tentatively approach the corn, realize what it is, and eagerly gnaw one cob apiece. What about the utility apples I was saving for a pie? Filling my pockets and hands with apples, I creep outside once more. As I toss apples to her, the largest female regards me solidly, eye to eye. An apple lands about three feet in front of her, and still she watches me carefully, then walks toward the apple, slices it with one bite, and eats with a mixture of surprise and relish. Apples in April! She looks back at me, allows me to settle low on my haunches and watch her and her family. I try not to move.

In time, she wanders toward the others, also happily eating apples and corn. The deer will survive at least one more day because of this food, maybe a few days, maybe long enough to get to the next decent meal. Knowing that, my heart lightens. It is a moment sealed in a glass paperweight, a scene to be reflected in a gazing ball, a time of peaceful communion with nature. And there I sit on the grass until evening drops a gray screen over the air and daylight drains away. At last the deer become startled by something real or imaginary and trot back toward the woods, with the largest doe leading the way along the fence. When she finds a place she feels comfortable with, she lines up squarely and hurdles it. The others pace nervously. One stands before the fence, lifts a foot as if to jump, thinks about it again, backs up, paces, then once more aborts the attempt and finally risks it: a five-foot jump straight up. Her hooves graze the top rail as she clears it. Soon the others follow, launching themselves from the tidy world of humans back into their familiar pandemonium of green.

3

The spring garden wakes slowly, like a sleeping village. Despite five days of rain and little sun, new plants raise their colors every day. Some continue to doze, suspended between seasons, others unfurl overnight. Transpose this annual drama into cemetery terms—thousands of zombies waking from their winter graves—and it wouldn't sound so cozy. Change the term "weeding" to "murdering" and spring cleanup wouldn't sound like a harmless activity for church ladies, especially if you saw them stabbing the dirt wildly with their trowels, ripping out one dandelion after another, a look of "Take that, you swine!" satisfaction in their eyes. Each day brings new discoveries. It's like finding coins in the grass: purple dwarf irises, white snowdrops, tiny fuchsia anemones, delicate purple-and-green fritillarias. The temperature spins like a weather vane. Robert Frost captured well the atmospheric indecision of April in his poem "Two Tramps in Mudtime."

The sun was warm but the wind was chill.
You know how it is with an April day
When the sun is out and the wind is still,
You're one month on in the middle of May.
But if you so much as dare to speak,
A cloud comes over the sunlit arch,
A wind comes off a frozen peak,
And you're two months back in the middle of March.

Suddenly the world grows noisier. Blackbirds are the first migrants to return each year, sometimes flying in mixed clouds of grackles, redwings, cowbirds, and rusties. This year there are thousands of red-winged blackbirds. When I go biking I'm surrounded by three of their distinctive calls: a short liquid sighing, a chat-chat, a long buzzer. I make a game of spotting them perched on cattails, telephone wires, fences, and branches. The males arrive first, to duke it out for territory (about a quarter-acre each), where they'll settle and lure a harem of quietly colored females several weeks later. They are birds that live in the margins, along marshes and meadows. For ten miles down country roads, I hear them warbling as loud as cicadas. Flashing red epaulets, the males make a territorial call birding books describe as "Oaklaree." But to me it sounds mechanical, like a tightly wound spring unwinding or an urgent buzzer. Halfway up a long hill, I coast back to where a large flock is buzzing from a canopy of oak trees, and circle a few times, letting my ears fill with the liquid of their call.

After blackbirds, the bluebirds and robins appear, the song sparrows, flickers, a dozen species of waterfowl, fifteen species of raptors, including red-tailed hawks, which sometimes visit my garden to grab a four-legged lunch. Mallards start courting quacksomely in the shallow water atop the pool cover. A dawn chorus signals the start of the breeding season—an open-air den of iniquity to some, I suppose, as grackles copulate and dwarf irises offer their sexual parts to passing insects. You'd think all the Howdy do-ing, nasty threats, and simple boasting would sound raucous, a sort of symphony played in reverse. Yet it doesn't. Somehow the different calls blend like a kindergarten band of sandblocks, pipes, kazoos, drums, and bells. Each morning at sunrise they start on cue.

Jeepers, creepers, have you heard those peepers? T. S. Eliot claimed April is "the cruelest month," but it's also the croakiest. After a long winter's hibernation, frog species bestir themselves in March and April, and rising from pond mud and leaf litter, they loudly proclaim their horniness. Archie Carr, a naturalist who spent his life studying frogs, writes in *The Windward Road*: "I like the look of frogs, and their outlook, and especially the way they get together in wet places on warm nights and sing about sex."

Are the frogs singing? Depends on your taste for bagpipe music. Whirring, trilling, croaking, plunking, twanging, whistling, quacking, snoring—all come from inflating an air sac on the cheek, under the chin, or in the armpit. Yes, under the armpit, like high school boys making rude sounds. My personal favorite is the green frog plunking on one limp banjo string. But there's something otherworldly, a thrill like no other, when you hear the loud bell-like jingling of tiny tan tree frogs. People call them "spring peepers," but I swear those pipsqeaks have microphones. Only one inch long, a peeper can distend a throat pouch to nearly that size, and use it as an amplifier. When they solicit females in unison, their appeals carry a mile or more, drowning out car engines and stereos. Forget conversation. Think thousands of loud, boastful, sex-starved males on the beach over spring break. Many people hear them, but few see them. These small brown frogs have a distinctive X pattern on their backs. They climb trees and feed on insects as many frogs do, but much of their life is a mystery. Where do they mate? Where do they spend the winter? We pay attention to them only when they're onstage and operatic.

Then there are the wood frogs and American toads, which I always picture as rowdy frat boys throwing an orgy. When the females arrive, the males try to mate with anyone, anywhere, in any position. Latching onto the females while mating, they violently kick away rivals at the same time, and you can imagine the chaos. In this mêlée, males often mount other males and try to mate, so for efficiency the species has an "Oops, you made a mistake, try another frog" sound, which scientists politely refer to as a "release call." Archie Carr was right about frogs. I've never actually heard one say "gribbitz," but I understand California tree frogs do, hence their popularity in movies. According to biologist Adrian Forsyth, who wrote classic books about the rain forest, we like frogs because they remind us of ourselves, which is why so many cultures have myths in which frogs become people.

> [They] have a way of facing you with a goggle-eyed gaze that is disconcertingly humanoid. Their huge wrap-around mouths, while perhaps not actually resembling a smile, are certainly not frowning. Sitting hunched up as though in anticipation, they assume the

> posture and calm demeanor of patient listeners ready to participate in conversation. . . . Our emotional responses to these appealing features of frogs are not irrelevant. The art of natural history lies in allowing such personal reactions to organisms to lead us into their biology.

Never hide a frog in your mouth. Never lick a toad. Never kiss a warty small green male, however princely. Disgust is an underrated strategy. Many toads exude a toxic slime that makes predators recoil. The poisons tend to be hallucinogens, which teenagers are often tempted to sample, so each year some die from toad-licking. Toads won't give you warts, but they can kill you.

Little boys can be notoriously cruel to frogs and toads. A British friend told me how, when he was ten, he and his pals used to find frogs, put straws up their rear ends, and inflate them until they burst. "Why would children do that to a living thing?" I asked in anguish, and he answered, "Because we could." It pained me to hear, but I do appreciate that nature is "red in tooth and claw," as Tennyson says. For example, skunks eat toads, but first torment them, rolling them hard until they exude all their poison. As for the skunks, their passion for Yoplait yogurt has caused the company to redesign its containers. Apparently, skunks go through the trash in search of discarded Yoplait containers, stick their snouts in to lick the delicious leftovers, and get stuck with the containers on their noses. Animal protection organizations complained, and Yoplait redesigned their packaging with a skunk-friendly device that makes the cartons easier to remove.

Spring also means the return of two of my favorite people, Chrys and Bill, who are gardeners of a special stripe: passionate about nature, knowledgeable, and devoutly organic. They're not guests in the garden, but a welcome part of its ecology. In a large, thriving garden, no great deed is accomplished alone. They do the heavy work, or whatever I don't especially enjoy. Anything that bothers my dodgy knee qualifies in both categories. This might include planting a tree, applying sulfur to control black spot, creating a new bed, burying platoons of spring bulbs, spreading a mountain of mulch, clearing thickets. I actually enjoy repetitive chores like the endless deadheading of minute aster flowers, but I'm not keen, say, on picking slugs off by hand. Roses require lots of pampering, and I have a hundred or so rosebushes. So one day a week, Chrys and Bill garden for me, and that means on the other days I can spend an hour tending and preening—removing aphids and black spot leaves by hand, deadheading, fertilizing, attending to the garden's many unspoken needs. I also allow plenty of time each day for fussing with and admiring any flowers that may have opened overnight, and each day I like to make fresh flower arrangements, however small, for indoors. I collect vases; in the beginning I chose them solely for beauty, but now I picture the flowers that might rest there. I consider the size of the throat, for instance. It takes plenty of flowers to fill even a medium-throated vase. Some vases are too tall for regular use, and others not quite tall enough. Several large decorative vases look pretty on a high shelf, where they await the occasional bouquet of gladioli or tulips. Close at hand, I keep a variety of medium to small vases, unusual bud vases, and tiny vases that can hold only a lapel's worth of flowers. I also look for vases cleverly sculpted or richly colored. Part of the treat each morning, after gathering the day's roses, is choosing the vases that suit the flowers and my mood. Dancing naked ladies embossed around a green flute of Czechoslovakian glass? Perfect for a cloud of miniature pink Fairy roses, with tall twirls of curly mint in the center.

Today Bill has been laboring for hours with massive railroad ties weighing 100 pounds each, using a bow saw to build a new two-tiered rose garden. Chrys and Bill have created a life reminiscent of the sixties and seventies, a return to simple things, the home-grown, the handmade, home-schooling for their daughter, physical labor, affection for the land. It's not surprising that Chrys has a large tattoo of a columbine, her favorite flower, on her stomach, or that they've built their own energy-efficient house by hand, or that

Chrys has legally changed her name to "Chrys Gardener." Her tattoo is of an *American* columbine. In my garden I have many varieties of columbine, some European, some American. The American columbine has spurs sticking out and looks a little like a jester's hat. But there are no hummingbirds in Europe, or the entire Eastern Hemisphere for that matter, so columbine didn't need to evolve long spurs. Instead they mainly grew round and ruffly in fabulous colors, like darkest plum and puce. Fortunately, the American and European columbine have interbred in gardens to produce colorful, round ruffly flowers with spurs, the best of Old and New Worlds.

I try to balance my aesthetics with Chrys's and Bill's. For example, in spring, I prefer parrot tulips with splashy colors and frilly petals. But Chrys loves deep saturated tones, and I let her choose the new tulips each year. I love all tulips and am not finicky enough about them to stifle her. I suppose this is similar to the marriage wisdom of "Choose your fight." Some things are worth squabbling over in a marriage, others really aren't. So, too, in a garden's extended family. For instance, it would be quicker if Bill used a chain saw on the thick railroad ties, but I know he prefers the quiet beauty of the bow saw, which appeals to his sense of touch, sound, and honest sweat. As long as the saw stays sharp, it works well, though slowly, but the intense outdoor labor of the day pleases Bill. It may add to his intimate feel for nature. It may even be a question of tact. I wouldn't want to interfere with that for the sake of efficiency.

I'd like some new delphiniums to replace the small grove of petaled towers devoured by slugs, and if I went to a nearby nursery I'd find a glorious selection of colors to choose from. But Chrys has raised delphiniums from seed, and she's understandably proud of their growth; so I'm buying hers instead and letting her choose the colors. I did suggest a Knights of the Round Table theme—ones called King Arthur, Black Knight, Astolat, and so on, knowing the range of colors that might include.

"Please plant that large crab apple tree at the foot of the garden," I might ask Bill, thinking how its radiant red leaves and pink flowers will shine next spring, and also how the deer will enjoy the fruit. But I don't specify exactly where to plant it, because Bill and Chrys cherish their own sense of balance and order.

I experience my garden as personal and private, but it also extends into the lives of many others. Somehow its jigsaw puzzle grows, piece by piece, and everyone must be content: the stone mason; the tree surgeon; the drainage man rerouting an underground creek; the landscape architect; Chrys and Bill; my Paul, otherwise known as Paul West, author of forty-some smart and stylish books, primarily novels. Paul functions mainly as statuary in the garden, surveying things from a chair on the patio while listening to classical music and dreaming up a beautiful new book filled with other realities, other worlds. Although he appears to be a reclining bronze, he's actually prowling the wharves of imagination, where any roustabout idea may turn to honest labor. Sometimes he swims or plays whistle games with the birds, and his memory of an English childhood requires a lawn—just in case he wants to set up a cricket pitch. I'm not opposed to lawns. Lawns provide a pool of calm in the midst of flowery commotion. The Japanese call it *ma,* the space between. In flower arranging the empty spaces are vital to the total effect. Emptiness isn't always negative. It allows one room to act and concentrate, it forms and defines. In Istanbul, domed mosques carve the skyline because of the negative space between them. Books need transitions; the sun can't always be at noon in them. Pauses are meaningful in poems. In haiku, for example, silence and suggestion are as important as statement. In gardens, they set up the moment to come. A pause in the rhythm of the garden makes the rhythm clearer, and gives the narrative a before and after. I try to balance my curving flowerbeds with Paul's expanse of neatly mown green. Everyone's view of the same garden is different, just as every sibling has a unique view of his or her family, and that's fine with me. Nature always has the final word, anyway.

4

Spring has hit with a visual thunderclap followed by trumpet flourishes. Dazzling yellow forsythia bushes zoom all over town. Some people have planted long forsythia hedges, a clue to one's personality. First bush to bloom, forsythias gush color, their limbs flail in the wind as if on fire, and several bushes can light up a whole street. They startle in early spring by bringing clouds of gold to an overcast world. But when their flowers fade and drop, they return to an existence so ordinary, so timid, so drab, that they're easy to confuse with weed bushes best described as "brush." Craving a fix of color after a hard gray winter, some gardeners plant forsythia prominently around a house, no matter how fleeting the blooms and how boring the bushes for most of the year. So forsythia really appeals to impatient, easily inflamed souls like me.

It must also have appealed to the legendary collector of exotic plants and professional gadabout Robert Fortune. After the Treaty of Nanking opened up China to the West in 1842, thirty-year-old Fortune relished the idea of plant collecting on a grand scale in a nearly unknown land. The British Horticultural Society offered him the opportunity he craved. For one year, he would travel throughout China, collecting unusual plants and seeds and learning about Chinese horticulture. For this, he would receive £100. It wasn't much of a salary, and it included strict guidelines for how he was to travel and what sorts of plants to gather. The society was especially keen for him to find new teas, blue peonies, double yellow roses, azaleas, lilies, and various fruits. Both he and the society knew, from the accounts of a few travelers, that danger would be a relentless companion. They equipped him with a Chinese dictionary, collecting tools, airtight glass cases (a sort of portable greenhouse), and a lead-lined stick they referred to as his "life preserver." In time he persuaded them to include a shotgun and some pistols, provided he agreed to sell them on his return and give them back the money.

On February 26, 1843, he set sail, arriving in Hong Kong four months later. Before his Chinese rambles ended, he was waylayed by pirates, shipwrecked during a monsoon, nearly trapped in a wild boar pit, felled by sunstroke, and repeatedly robbed and beaten. Food was scarce. Dressed in Chinese clothes and wearing a false pigtail, he stole into the then-forbidden regions of the interior. Traveling illegally, he couldn't risk eating at inns where his accent and his awkwardness with chopsticks might betray him. His journals tell of a hundred hardships and misadventures. In Shanghai, he records that locals called him Kwei-tsz, which meant "Devil's child." Wherever he journeyed, he was an oddity greeted with suspicion, hostility, or outright violence.

That he was immensely brave and stoic goes without saying. What really led to his ultimate success was superhuman stores of patience and persistence; a fascination with new customs and ideas; fondness for ordeal; and a passionate obsession with plants, which he seemed to regard as earthly angels or tiny arks of the covenant that he must pursue and preserve, at whatever cost. In time, he won the trust of local gardeners and officials, and spent nineteen years exploring China and Japan, both of which he came to love. I don't know if he ever paid the society back for the guns stolen by highwaymen. Or, for that matter, what his private, off-duty life was like (he wrote many letters, but all have been lost).

Fortune sent home a wealth of flowers, including some gardeners' favorites—forsythia, bleeding heart, tree peonies, jasmine, anemone, weigela, rhododendron—which were first bred and cultivated by the Chinese. It's a pity those flowers no longer remind us of Fortune's wild adventures in China. Perhaps forsythia should be named Yellow Fortune or Fortune's Fool, and bleeding heart named Fortune's Breeches, or Lady Fortune, Bathing.

No, its exuberant yellow was named after the exuberant Scottish gardener William Forsyth, director of the Chelsea Physic Garden. For a while, forsythia bushes were all the rage in England,

and then they slowly lost their novelty and grew unfashionable, even vulgar. Forsythia was just too easy, too gaudy, a reliable scream of color in any neighborhood or rubble-strewn lot, offering itself cheaply to passersby. When it became "common," as the class-conscious English also said of vulgar women, it lost its allure.

This season the narcissus are staggeringly strange and wonderful. I'm glad I planted extra varieties last year. One of my favorite new ones has ruffled yellow petals with an orange interior and a gold stripe down the center of each petal. The eye jumps from whole flower to stripes, foreground and background, over and over. Are they daffodils or narcissi? All daffodils are narcissi, but not all narcissi are daffodils. I say narcissus or daffodil interchangeably, depending on my mood, but, technically, they differ according to number of flowers per stem and length of trumpet.

Part of my excitement is not knowing what's coming up where. In the fall I gave bags of bulbs to neighbor children and said: "I'm going indoors. Don't let me see where you're planting them. . . . Plant them absolutely anywhere you like." I was curious how outlandish the children might get. Would they plant bulbs under a rock? in a gutter? deep in the woods? Instructed in the how, not the where, they might even try planting them in hollow tree stumps. Wouldn't that be grand? Aerial daffodils growing like orchids. Thus far, I see them blooming predictably inside the beds, edging the woods, flanking the mailbox, clustered beneath the apple trees.

I've never heard of anyone dying from a meal of daffodils, but they are poisonous, and I'm glad, because no creature will eat them, not even deer. The sap contains sharp crystals of calcium oxalate, an irritant that also seems to bother other flowers, which wilt if they share a vase with daffodils. A truly deer-proof flower? I've planted thousands—possibly 6,000 at this point. I've chosen species that bloom at different times, and even though last week's unexpected snowfall suffocated many, this morning a battalion of beauties has opened, guzzling the sun, and thousands remain in bud. Staggering blooms is a cardinal rule. My goal is an endless procession of daffodils, from forsythia bloom right through tulips.

The tulips traveled as far as the forsythia to reach my garden. Even the name, tulip, comes from the Persian *dulband*, or "turban." Not because they're turban-shaped, but because Turkish men used to tuck a tulip into their turbans. Picture a man with a tulip sticking out of his turban. No stranger than a woman with a peacock feather dangling from her hat. Just as rare feathers became prized by milliners (and led to the extinction of certain bird species), unusual tulips fetched high prices. They entered Europe in the sixteenth century from the court of the Turkish sultan Suleiman the Magnificent and quickly rose to the status of royal pets in Dutch and English gardens. In Europe, they took the names of nobility: Alexander the Great, Duke of Marlborough, Artaxerxes, and Black Prince. They served as currency during the "tulipomania" of Holland, when financiers speculated on tulip futures, and rare bulbs were traded for homes, carriages, fortunes. Owners protected their precious bulbs from disease, theft, and cold. There's at least one account of a man who froze to death because he used his last blanket to cover his garden of tulips.

Striped tulips fetched the highest prices, in part because they couldn't be grown on purpose. They were a petaled mystery. Growers refer to them as "broken," which sounds bad, but the implication is only that the color is broken into stripes, not that the plant is injured. Yet, in a sense, it is; although no one knew it in the heyday of tulipomania, the stripes are caused by a virus carried by aphids. When the Dutch tulip market crashed in 1637, fortunes and lives dissolved. In time, tulip speculation became illegal, and Holland's romance with tulips transformed its economy and image as a land of flowers.

Red-and-white-striped tulips swaying in the breeze remind me of the peppermint starfish I once saw on a coral reef. There is something truly oceanic about a garden's waves of color and wind. As I admire yellow pointy-petaled tulips, bright red peony tulips, and large apricot and orange tulips, a shark-shaped shadow appears at my edge of vision. A crow lands on the freshly graded lawn and starts eating the grass seeds. I whistle it away.

Last winter, I raised the wire fence around the backyard to keep the deer from the tulips and roses, and my, my, did it work. A crowd of turban-headed tulips is rising like a fantasy bazaar. Every day new stripes mingle with new pastels and screaming oranges jostle reds. Browsing deer stare through the fence at the overflowing paradise only inches away.

On the first rose patrol of the season, I discover the casualties of winter. Most years, there's death and mutilation in the trenches. This year, only seven roses went the way of all canes—John Kennedy, Stainless Steel, Signature, Candelabra, Tropicana, Platinum, Angel Face—though some, like the yellow-and-red-striped George Burns and the purple-and-white-striped Purple Tiger, have returned in a weakened state. The melting and freezing spells obviously clobbered them. A climbing Don Juan has no leaves or buds. I transplanted it from the front yard last year, and it doesn't take much to send a Don Juan into shock.

The redbud is an encrustation of purple flowers. The woods have begun leafing up into the thick green veil I look forward to all winter. The raised bed in the front yard is especially beautiful with blankets of creeping phlox spilling over the railroad ties: lavender, pink, and pink-and-white-stripe. Yellow and orange crown imperials (Fritillaria imperialis) look like dancing cranes. Dwarf purple iris petals shimmer iridescently in the sun, magenta to blue purple. Tufts of yellow euphorbias and pools of cream-and-white narcissus complement the purple. I laugh—each iris has a comb that's a distinct, pudendum-like stripe of fur.

I've started building up a new bed with aged horse manure. I use aged manure because it would be hard explaining a fresh street pile of horseshit to my neighbors, no matter how I laundered its name. After two years, it doesn't smell, nor does it waste its nitrogen in decomposing, making it available to the plants instead.

A beautiful andromeda bush, with cascading flowers like small white vertebrae, now sits below one of Paul's study windows. A purple-leafed smoke bush grows below his second window, and at the corner of the house two red-leafed sandcherries provide a shower of pink flowers. On the west side of the house, viewable from my bay window perch, an allspice displays strange nutlike brown flowers. Two climbing clematis will bloom blue, I hope. And two dwarf fothergillas are covered in white bottlebrush flowers. On the other side of the bay window a rose of Sharon now grows all alone, but I plan on adding an azalea or rhododendron for company.

A half-circle of purple crocuses with saffron-rich stamens lead to where a hellebore is supposed to be. Its green flowers delicately speckled with dusty rose herald spring each year, but tomorrow is Easter Sunday and there's no sign of it. I spot a tiny green hand with pursed fingers poking through the dirt—could that be the hellebore weeks behind schedule? Maybe so. The greenhouse effect scrambled the weather patterns. Two years ago temperatures ran so high that upstate New York entered a different climate zone, and some annuals actually overwintered. Last winter supplied early blizzards, before people could protect their roses with leaves, pine branches, or burlap. But that light blanket of snow stayed for months, insulating the beds better than anything humans might have planned. It also brought an unwelcome immigrant.

Today the first case of locally acquired Lyme disease was reported. A girl walked into a doctors's office with a target-pattern bite on her leg that tested positive. A few townsfolk have had Lyme disease, but they picked it up elsewhere. Now it's officially a local problem. Our changing climate brings mixed blessings: fabulous gardens, little snow to shovel off the roof, lower heating bills, fewer blizzard accidents. But cold winters always kept us safe from the Lyme tick, some part of whose life cycle found our winters deadly. A similar situation exists in the Hamptons with cockroaches. So many Manhattanites travel regularly to the Hamptons you'd think cockroaches would be constant houseguests, but some element of the Hamptons' climate—the saltiness? the sandy soil?—is inhospitable to them. Lyme ticks thrive there, however; 12,000 to 17,000 people contract Lyme disease each year. It's the United States's malaria: the most common insect-borne disease.

It's most often spread by the white-footed mouse, a little typhoid Mary, and the consequences can be ravaging. Another good reason for biodiversity. A large population of small mammals and lizards keeps the ticks in check, and that may be our secret weapon—lots of low life. Biodiversity also means a wealth of plants, and one never knows where the next heart or cancer drug may come from. Recently a local graduate student discovered a new broad-spectrum antibiotic while hunting for fungus in the woods. A miracle drug may be hiding in plain sight in someone's garden.

5

The leaves of the faded daffodils need to bask a while to gather sunlight, nourishment for next year's flowers, but they're unkempt and take up precious space. So I bend them over and wrap one leaf around each folded stalk. "Cue the Disney music," I say to myself, as I create chorus lines of green dolls. Gardeners give many names to the wrapped narcissus leaves. I've heard them called "sushi," "mummies," "spring knots," or "Zen warriors." Myself, I prefer "leaf dollies." Best to have a cozy name when you realize you have only 3,000 more dollies to fold. Devising a vocabulary for gardening is like devising a vocabulary for sex. There are the correct Latin names, but most people invent euphemisms. Those who refer to plants by Latin name are considered more expert, if a little pedantic.

One day, as I was biking by a corner yard at the base of a hill, I exchanged niceties with the owner and told her how often I've admired her garden, sometimes pausing beside her small pond to watch the waterlilies and listen to the banjo-string-plucking frogs. She invited me beyond the gate to look at the grounds. Not a picking garden, it's planted in mounds, as much for foliage as for flowers. After all, she explained, when the flowers are past, you still have to live with the foliage, and that should be attractive. Her false indigo was covered with seedpods, which meant that she didn't bother to pick any. Nothing wrong with that. Traditional Chinese gardens aren't meant for picking. I end up with only two or three pods left on my two false indigos, because I bring in nearly all the tall batons of sweet pea-like blue flowers. Why didn't I plant them alongside hollyhocks and gladioli, like a small armada of tall ships? I envied her her large smoke bush, which seemed to be thriving. Guiding me along her garden paths, she referred to each plant by Latin name, which some gardeners do a little fetishistically. It's not that they don't know the common names, and using such esoterica puts a hedge between themselves and the rest of the less knowledgeable world. I find it more picturesque to say "lamb's ears" or "turtlehead" or "monkshood" or "love lies bleeding." I use Latin and Greek names only when necessary, for the sake of clarity. *Ligularia*, for instance. I have two different versions of this dramatic, broad-leafed flower in the garden. The Rocket (*Ligularia przewalskii*) forms two or more spectacular batons whose buds begin opening from the bottom up, making a slow-motion skyrocket of blazing yellow. Its cousin, Big Leaf Golden Ray (*Ligularia dentata*)—which, just to confuse things, is sometimes known as Othello—forms frilly orange flowers atop gawky limbs and looks for all the world like a crested African bird. Because the *Ligularias* look so different, I use both Latin and common names at nurseries. I enjoy the mild hallucination of saying The Rocket and Big Leaf Golden Ray, which sound like monikers in a *film noir*.

"Beware of Shakespeare!" a man said to me recently. "You can't trust any of the characters. Othello is charming, but too fruity. Pretty Jessica is dependable, but downright common. Jacquenetta is appealing, all right, a real buxom country wench, but completely unstable. Prospero can be subtle, with an interesting spectrum of moods, but just doesn't appear for long enough. Proud Titania, when you come down to it, has too many problems to keep track of. For my money,

William Shakespeare is a gaudy giant, but requires altogether too much work."

That did it. "Listen," I said, "I think they're all brilliant creations. Okay, you don't get much of Prospero, but what you do get is rich and unforgettable. Sweet Juliet has the sort of blushing extravagance I would defend to the grave!"

We were two gardeners talking about roses. Wouldn't Shakespeare be surprised to find many of his characters transmogrified in gardens all over the world? He'd probably hate the bad press the David Austin English roses with Shakespearean names are getting at the moment. And how confusing to have both a rose and a ligularia called Othello.

At nurseries, salespeople want to know which one you mean, so knowing the Latin name can be useful, especially if you've graduated without noticing it from casual hobbyist to pursuer of rare and unusual varieties.

Bill phoned early yesterday morning to say that he and Chrys had an emergency and would come a day later to help bind up the leaf dollies. When they arrived this morning, Chrys looked troubled. She explained that a client of hers, an older woman whose garden they had planted many years ago and continue to maintain, telephoned in some distress. The lady was moving to Florida in three weeks and she had given her home to her son and daughter-in-law and their new baby. The young couple decided that they didn't want a garden but a nice flat lawn for their child to run around on. So they phoned a landscape company and instructed it to tear out the entire garden—flowers, paving stones, ornamentals, everything—and just dump it in the woods. Anger welled up in Chrys as she described the face of the woman watching her beloved garden full of memories being destroyed. Chrys of the ponytail, columbine tattoo, and gentle ways. That any son could be so selfish seemed unimaginable.

"He couldn't wait three weeks until his mother left?" she said angrily.

"What on earth did you do?" I asked.

Chrys explained that she and Bill had salvaged the shade-loving plants and settled them in the woods behind the house. The rose-bushes and other flowers they loaded into their truck. Then Chrys had tried to comfort the woman by promising to plant exactly the same garden, adding that whenever she came north she could visit it and see it blooming.

"Imagine," Chrys said in anguish, "he told the landscape company to rip all the flowers out—even the tea roses—and dump them all in a trash pile in the woods. He did this while his mother watched!"

Why not drive a knife into her heart? It was worse than putting her out to pasture. It was pure Greek tragedy—a son uprooting the most vital part of his mother's life, the blooming part that she had been cultivating for years. Why did he feel the need to destroy her perfectly nurtured little world, one of her last sources of pleasure and accomplishment, her belief in a future she would live to see? Why do so in front of her, if not as an expression of his power? It's hard to guess what revenge drama he might have been staging.

This is not the first time I've heard of a garden being used as a weapon. A neighbor's husband divorced her for a younger woman, and also received their house in the divorce settlement. Knowing that he would soon be moving in with his new wife, my neighbor methodically ripped up the garden's many beds of rare and expensive perennials and replaced them with annuals she alone might enjoy before she moved out. The newlyweds could arrive to a dead garden. For all I know, she may have poisoned the dirt as well.

Violence in the garden? Thanks to the ancients, one can find nonstop murder and mayhem in a spring garden. For example, narcissus commemorates the abduction of a beautiful princess. According to one version of the Greek myth, Zeus created the flower to help out his brother, the lord of the underworld, who was in love with Persephone, Demeter's daughter. One day, Persephone was gathering flowers with her friends, when she spied a brilliant blossom across the meadow. Her friends hadn't seen it, and she laughed as she ran to discover just what it was. She had never seen one so radiant before, with so many flowers bursting from the roots, and a seductive fragrance, both sweet and animal. Just as she reached out a hand to caress it, the earth yawned open at her feet, and "out of it coal-black horses sprang, drawing a chariot and driven by one who had a look of dark splendor, majestic and beautiful and terrible." He grabbed her and held her tight, and galloped away with her to his world of the dead, far from the sunlit joys of springtime.

Leucothoe, a squat bushy shrub outside my window, was named after another princess, a Persian woman whose jealous husband chased her off a cliff and into the surging ocean below. Apollo took a fancy to her and changed her into a sea goddess, and, when he tired of her frothy ways, into a sweet-smelling plant.

The bright red anemones that bloom in the summer take their name from Adonis, who was out hunting one day when a wounded boar turned and gored him in the groin with a tusk, castrating him. It was an excruciating and deadly wound. By the time his lover, Aphrodite, found him, he was delirious and nearly dead. Weeping, clinging to him, she moaned:

Kiss me yet once again, the last, long kiss,
Until I draw your soul within my lips
And drink down all your love.

But by then he was far from her words or tears, down in the underworld where she could no longer reach him. As each drop of his blood fell to the ground, delicate red flowers bloomed. His severed penis was said to have run off and become his son, the erotic god Priapus.

A meander of purple, apricot, and plum hyacinths is sprouting along several pathways. Hyacinths get their name from a young sweetheart of Apollo's, a boy he accidentally killed. The two were having a friendly discus-throwing match when Zephyr, the west wind, whose attentions the boy had earlier spurned, got into a jealous rage and blew on Apollo's hand so that it slipped. The discus flew off at a freak angle and broke the boy's neck. Horrified, Apollo pressed him to his heart and wept. As the boy's blood trickled onto the grass, a single phallic flower grew as his monument, a beautiful purple column whose petals spell the letters that form the Greek word for "Alas."

Most anywhere you turn in a garden, nature tells painful stories of lost beauty and innocence, lost love and life. There was so much violence and loss in Greek myths that they liked to turn gods and mortals into flowers. It solved a lot of problems created by prolonged anguish; it provided an agreeable form of reincarnation; and it lifted the bloody tales into a slightly sunnier mood of redemption. A beautiful young woman died of a broken heart in a cave? Pitiful. A handsome young man drowned? Tragic. But at least they continue to live in nature, eternal and much-loved, blossoming each spring. Not everyone thinks of Apollo's broken-necked lover when strolling beside hyacinths, nor for that matter of penises growing in the garden, but if you listen hard you can hear the wind whisper "Alas."

6

May Day. I may believe in the rebirth of the garden. I may visit several nurseries where tractor trailers are delivering the year's first roses, pansies, and other gems for Mother's Day. I may go cycling while woodpeckers rat-a-tat-tat loud as gunfire. I may find just-picked baby greens and parsnips at the farmer's market. I may shmooze by the lakeshore with growers, carvers, clog dancers, and potters as I eat purple Laotian rice covered with Thai vegetables and a pyramid of cilantro. Later I may weed between the patio rocks for the first time this season, using a spade and a firestick. I may contemplate the return of the mayfly to Lake Erie. I may lie in my bay window and enjoy the magnolia, whose blossoms will drop at the first hard rain. Such beauty, such impermanence, and no foreknowledge of its life span adds a poignancy more animal than botanical. This Chinese magnolia has large pink flowers that billow like sails. In a light breeze the whole tree quivers. I may lie in the bay window every single day the petals are open. Waxy white with a pink throat and base, they look like delicate hankies held by invisible hands.

Many people, from architect to builders, have tried to persuade me to cut down this stately magnolia, because it occupies valuable space where the house might extend. But the magnolia is a beacon at the end of the street, the heart of spring, even though its petals only linger two weeks or so. In that time, it's as exquisite as a Zen drawing of snowfall trapped in a tree. The pictograph for "leisure" is the moon captured among tree branches. What might snowfall in a magnolia tree represent? Old age? Tolerance? I couldn't bear to lose it. A large tree with eight main limbs sprouting from the trunk, and an umbrella of twenty-five feet, this magnolia may be forty years old. It's an ancient breed. Earth's first flowers bloomed on magnolia-like plants. Goblet-shaped blooms just like these have existed for 200 million years. What is a human life span compared to that? Break open the finger-shaped seedpod and you find a perfect starfish filled with many seeds. It's a wonder the world isn't sea-to-sea magnolias.

One day I watched a squirrel twist off a large magnolia bud, toss it to the ground, and lick sap from the remaining twig. The sap would carry minerals on the way up and nutrients on the way down, a good tonic for a squirrel. The tough layer of bark is supposed to protect the sap, but insects, fungi, birds, and mammals have all devised clever ways to plunder a tree. Of course, trees get their own back by trapping insects in sap; the fossilized form of those drops we prize as amber. This was the first time I had seen a squirrel twisting off a magnolia blossom as if it were a bottle cap, but I have watched hungry squirrels peel thick bark in winter. They can be sap vampires.

The sycamore in my front yard sheds its bark every year. The garter snakes shed skin, too. The deer and squirrels shed winter coats. The quaking aspens are always peeling thin papyrus-like skin. The top layer of our skin sheds every two weeks. I sometimes feel surrounded by plants and animals sloughing off their previous selves. We mine bark (paper birches for canoes, cork oaks for bottle corks) and eat the perfumed bark of Sri Lankan saplings (known as cinnamon). I've seen birdhouses created from the bark of fallen trees. We weave bark fibers into flax, hemp, and ramie. We get tannins, gum, mucilage, rubber, and latex from bark, whose resins give us shellac, lacquers, frankincense, and myrrh. Essentially, bark provides the circulatory system for a tree, and yet it can vary enormously. Some bark develops thorns or spines, others dots or flakes, and many grow beautiful colors or patterns. One of the prettiest trees is the Australian lemon-scented gum tree, *Eucalyptus citriodora*, whose scaly bark is a reddish purple that cracks and peels to reveal a silky blue-gray trunk. There's caffeine in the bark of the Amazonian yoco vine (*Paullinia yoco*), and many medicines and poisons in the bark of other trees. One could spend a lifetime studying the secret world of bark—and some people do.

At the farmer's market today I bought three birdhouses crafted from hollowed-out gourds that have been dried, shellacked, and threaded with copper wire. One gourd, with a long elephant trunk curved for perching, hangs in the magnolia now, at just the right angle for me to watch its doorway. In eyeshot, about thirty feet away, I've hung a larger, plump-bellied gourd, which would be perfect for woodpeckers, nuthatches, or titmouses. Then, in the back yard, I've hung a small gourd wrenhouse with a long Modigliani neck from a branch of the redbud, a dozen feet from my study window.

Local tradition warns us not to plant annuals until fear of frost has passed. Most years we get snowfall as late as Mother's Day, which keeps us on our toes. On the other hand, today will soar into the 70s—summer weather, planting weather—and it's hard to resist waking the garden up with a good shake of fertilizer and new plants. So I've brought home a tall, mature crab apple tree, thickly in bud, which I'll plant at the bottom of the yard. The deer may enjoy its fruit, and I'll enjoy its radiant red blossoms. Also bought two pussy willows, four forsythias, and four budded roses. They need no planters; the earth will both thwart and sustain them. But what would a garden patio be without exotic planters for more delicate flowers? Architectural elements have always been a part of gardens. In addition to several black metal obelisks, for climbing roses to investigate with spiraling limbs, I've arranged a curiosity shop of plaster planters: dancing Greek ladies, cherubs wrestling goats, baby angels, the faces of demure women from the Victorian period. Artificially aged to look like marble, they greet patio visitors with overflowing flowers. I chose miniature red roses, bicolor geraniums in mottled purples and pinks, bright yellow baby dahlias, and miniature purple-and-red fuchsias.

The Japanese lilac has proven a disappointment. I planted it last year with much excitement. By early June it was covered with large white blooms, which became fragrant only when they aged and opened fully. Before then, I could put my nose into them and smell nothing. After then, I could stand a yard away or quite far downwind and be blanketed in thick smell. All over town the Japanese lilac trees are blooming right now, some forty feet tall. Only mine is barren. For some reason my smoke bush hasn't formed its eponymous clouds either. Few things are begun with as much hope as a garden, and it can disappoint in direct proportion to one's anticipation. The late Henry Mitchell, who wrote witty garden columns for *The Washington Post*, pointed out that "wherever humans garden magnificently there are magnificent failures," and insisted that "it is not nice to garden anywhere. Everywhere there are violent winds, startling once-per-five-centuries floods, unprecedented droughts, record-setting freezes, abusive and blasting heats never known before. There is no place, no garden, where these terrible things do not drive gardeners mad." Defiance is the key, so one may as well forget worry or dread. The worst will surely happen, as well as the best. Some flowers will grow, others die, but the garden will continue, the gardener will plant new dramas and mint new dreams.

Out for a garden stroll, I bump into neighbors who tell me an odd tale. A three-car family, they weren't using their Oldsmobile much over the winter and spring. Last week it started making strange sounds, and they took it in for servicing. There a mechanic discovered something remarkable under the hood: wiring gnawed through and hundreds of walnuts that a squirrel had stored for the winter. My neighbors asked the mechanic to put the nuts in a box, took them home and carefully placed them in a shed in the backyard, hoping the squirrel would accept a new pantry. Heaven knows what the squirrel made of the nuts vanishing from the car and magically appearing in the shed. But it didn't have much time for confusion because soon afterward my neighbor's son was driving the car when he heard weird sounds behind him. In his rearview mirror he saw a terrified red squirrel perched atop the back seat. Pulling over, he opened the car door and the squirrel dashed out and away, miles from home. When the boy returned, he told his parents that he had "relocated the red squirrel." I hope the little critter survives. I hope it wasn't a female with nursing pups at home. I think it may have been the expressive male, Red Rodney, whom I haven't seen around for a few days.

Sitting on a stump in the garden, I watch the antics of two courting rabbits, jousting and dancing like a scene out of *Alice in Wonderland*. One doesn't think of rabbits acting so medievally. A dozen slender alliums have risen tall, formed spear-shaped heads, and begun to rock in the wind. In this phase, they look nothing like the explosion of purple satellites they'll soon become, multisprocketed spheres atop barely visible stems. Fiddlehead ferns, starting to unroll their leaves, look like long rows of violin necks. I love the garden's many masks and personas. Living things tend to change unrecognizably as they grow. Who would deduce the dragonfly from the larva, the iris from the bud, the lawyer from the infant? Flora or fauna, we are all shape-shifters and magical reinventors. Life is really a plural noun, a caravan of selves.

Study deceit among plants and animals, and you'll quickly realize that we're not the only creatures who wear masks and play tricks. Most plants are pimps and thugs. Because they can't walk, flowers will do anything, no matter how lethal, extreme, or bizarre, to get other life-forms to perform sex for them. Hence flower petals shaped like the sex organs of female bees, so that courting male bees will unwittingly spread the flowers' genes instead of their own. The more I learn about plants, the more impressed I am by their accomplishments, many of which we think only humans have achieved: controlling the wind and other forces of nature; clever forms of defense; the ability to adapt to new stresses; the knack of seduction and persuasion; and, especially, an impressive gift for deceit, exploitation, and barter. Trees and flowers undergo chemical fluctuations—do they have their own version of moods?

This metamorphosis reminds me of a lovely actress and dancer I first glimpsed playing a Scottish schoolgirl in a production of *The Prime of Miss Jean Brodie*. A week later I saw her hurrying across campus, out of makeup, an average coed in blue jeans. Soon after that she appeared in my class as a smart young student with a winning personality and a fine sense of humor. Later in the semester, when I attended a dance concert she partly choreographed, I was fascinated to see the modest, freshly scrubbed self I knew transformed into a voluptuous coquette, who, with the rest of her dance troupe, cavorted astonishingly well at high energy for two hours straight. I hadn't guessed that version of her. I wonder what versions of the other students might surprise me. Shimmering between sun glare and shadow, they move slowly around the garden now, trying to find their places as recent ghosts of memory. I'm already missing those seventeen students more than I'd imagined.

A course on creativity in the arts and sciences, my class attracted academic misfits of an enchanting sort. Typically, a student might confess: "I'm in nuclear physics, but my real passion is for medieval Irish song." On Mondays, a composer or entomologist or some other deeply creative person visited to talk about his or her process, problem solving, and passion. On Wednesdays, I met with students in a discussion section where we explored creativity in a mental free-for-all and think tank only slightly more in my control than it wasn't.

In the small garden of that classroom, conversations had overgrown trails, canopies and understories, bedrock bottoms, and intractable murk. Because the students—sophomores to graduates—came from different disciplines and backgrounds, they combined with great variety. I mainly provided a safe place to delve and encouraged them to unearth new ideas and take risks. In addition to their usual assignments, each week they turned in pages from their journals, which I would not grade or comment on; and so I had the great privilege of eavesdropping on the conversations they conducted with themselves about the nature of creativity, but also about the terrain of their lives. I watched ideas roost in their minds, I read their professional gripes and dreams and learned about regrets, fears, longings, and affairs. For fourteen weeks, I followed the movement of their minds and was struck most of all by their powerful need to grow.

The growth urge is something that all life shares, but I find it most fascinating in humans, where the need to improve oneself, progress, achieve, stretch one's limits, and other variations on the livingness of life shows itself with such ingenuity. Most animals have strong life forces and will fight to thrive. But humans also struggle to grow in understanding and fulfillment, sometimes making great sacrifices. Not always rewarded by happiness or pleasure, this growth is a pursuit of knowledge for its own sake, a quest for insights with no need to exploit them. We grow up, we grow old, we grow in knowledge and experience until the end. Perhaps we sense that growth is life and if we stop growing we will die. But I think it's more likely that our evolution was so chancy that we've inherited a fierce instinct toward mental growth, one that will lead us across space to other solar systems someday. That we can devise our own destiny is something we take for granted, but how startling, how implausible it really is, that flesh and blood might evolve over the eons into creatures able to run their own show, control their own evolution, store information outside their bodies, create evolving cultures and ideas.

In class, we identified some elements essential to creativity: risk, perseverance, novel problem solving, disciplined spontaneity, the need to make exterior one's inner universe, openness to experience, luck, genetics, a willingness to react against the status quo, delight, mastery, the ability to live not only one's own life but also the life of one's time, childlike innocence guided by the sophistication of an adult, resourcefulness, a sense of spirituality, a mind of large general knowledge fascinated by particulars, passion, the useful application of obsession, a sacred place (abstract or physical), among many others. That all these qualities, and more, might combine to produce what we refer to as a moment of *inspiration* is one of the great mysteries and triumphs of mind. Our minds grow. Not because we always need them to but because it is their nature, just as it is the nature of flies to seek out open wounds and the nature of flowers to follow the sun. Can there be growth without stress and struggle? In one of my favorite books, *On Growth and Form,* the classicist, mathematician, and naturalist D'Arcy Wentworth Thompson writes:

> {It} is a very important physiological truth that a condition of *strain,* the result of a *stress,* is a direct stimulus to growth itself. . . . The soles of our boots wear thin, but the soles of our feet grow thick, the more we walk upon them: for it would seem that living cells are "stimulated" by pressure. . . . The kneading of dough is an analogous phenomenon. The viscosity and perhaps other properties of the stuff are affected by the strains to which we have submitted it, and may thus be said to depend not only on the nature of the substance but on its history.

Although he uses ingenious analogies, I don't think Thompson was thinking how a person's strength of character is formed by a similar history of stresses and strains. His genius is for analyzing biological processes mathematically, in concrete physical terms. It's just that when he reveals the underlying structures and processes all living things share, one naturally thinks about the human condition. A paradox is that we are living organisms made up of nonliving parts, both mind and matter. It makes his mathematical work intimately human and suggestive. Study the engineering of the backbone or the bone wings of the pelvis and you can learn truths about sex or fashion. Study the forces that spur a cell's growth and you can learn truths about bread or the mind. It's a poet's treasury. I don't know if Dylan Thomas had read Thompson's book before he wrote these lines, but they strike a pure Thompsonian chord:

The force that through the green fuse drives the flower
Drives my green age; that blasts the roots of trees
Is my destroyer.
And I am dumb to tell the crooked rose
My youth is bent by the same wintry fever.

My twelve-year-old neighbor stops by to ask if I'd like to see a secret robin's nest she has found. We stroll to the end of the street, where, hidden deep in a low yew, behind many branches, sits a nest with three two-day-old chicks. Their mouths are orange, their eyes unopened slits. The mother robin perches in a tree at a safe distance. Bright triangles of orange, the babies keep squawking for food. Their twig nest looks tidy, but baby birds produce their own diapers, fecal sacs that hold their waste products. The parent bird just picks up the sacs and flies off to dispose of them.

Over the next week I check on the chicks each day, watch them grow, fledge, become more alert, until one day I stare in at them only to find them staring earnestly back at me. The eye contact is jolting. Furthermore, they're almost too large for the nest. The next day they have flown. I knew they would leave, and yet the empty nest still surprises me. Expectation doesn't ease the sense of loss. To grow is to change, to change is to require different needs and habitats. Without change there would be no novelty, no surprise, and without those fraternal twins life would feel flat as a postage stamp.

7

"A Guest in the Garden"

Each year I watch the redbud form thousands of lilac-pink eggs before the leaves appear. Surely a giant frog has left them, to become tadpoles not blooms. But tiny globes quickly swell and polka-dot the branches with hot pink tufts. At one stage, they look like champagne grapes, and they are edible, either raw in a salad or fried as a vegetable (redbud is in the pea family and forms flat pods in the fall). Folk healers brewed the bark as a remedy for diarrhea. I don't think of it edibly or medicinally, but as a banquet for the eyes in spring: pink fireworks at the exact moment they're beginning to explode.

My redbud stumbled in slow motion long ago. For decades, it's been kneeling on the ground and waving its long arms in a posture I suppose some gardeners would find ungraceful. The trunk has split twice, and a marauding wisteria spiraled tightly around many of the branches. In one place, they entwine like hard, braided muscles. Although the tree looks precarious, kidnapped, and downright broken-boned, it returns to bloom heavily each year.

Covered with heart-shaped leaves, the redbud is decorative even when the flowers have faded. I first saw a redbud at Monticello, Thomas Jefferson's home and research garden. An experimental horticulturalist, who swapped seeds and plants with naturalists worldwide, Jefferson was our country's first serious gardener, a man who loved redbud trees and planted many around his house. His country house, I mean.

I remember a day in 1962, when John Kennedy fêted the country's Nobel Prize winners at the White House. First he sang their praises, then quipped: "I think this is the most extraordinary collection of talent, of human knowledge, that has ever been gathered together at the White House, with the possible exception of when Thomas Jefferson dined alone."

No one was insulted. How could they be? Not only was Jefferson an enthusiastic and savvy botanist, he was the greatest American architect of his day, a deft man of letters, a tireless inventor, tinkerer, and gadgeteer. A collector of dinosaur bones, he reconstructed the first mammoth. He was a student of the classics and a book maven who compiled two of the greatest private American libraries (the basis for the Library of Congress). In 1766 he started a garden journal, in which he recorded all the details of planting and harvesting, first blossoms, bird study, weather, varieties of various plants, the layout of the beds, and his many botanical experiments. Because he kept records faithfully for fifty-five years, the local garden club has been able to restore his heirloom gardens. He left them a guidebook.

A champion of exploration, Jefferson sponsored the Lewis and Clark expedition, choosing Meriwether Lewis because of Lewis's passion for discovering new plants, and even sending Lewis first to Philadelphia to hone his skills under the tutelage of renowned botanist Benjamin Barton. When Lewis and Clark returned, with full notebooks and seed chests, Jefferson planted some of the seeds in his own garden—no doubt one of his main motives for the expedition—and sent the rest to Philadelphia for cataloguing and study.

In his spare time, he was president. But his political life, as he often pointed out, was merely "circumstance." His real passion was gardening. His farm didn't just provide an income he badly needed (when he died, he was $100,000 in debt); he used it to focus his many inquiries about nature. The early years at Monticello with his first wife, Martha, who died after only ten years of marriage, and the seventeen years there of his retirement, surrounded by his daughters and a dozen grandchildren and his mistress, Sally Hemmings, were the happiest of his life, he claimed. He spent forty years building and rebuilding Monticello, planting and unearthing its gardens. Nothing in life thrilled him more, he said, than "putting up and tearing down." And there was a sense in which his presidential life in Washington felt cramped and suffocating. His galloping mind was harnessed there by the cares of a single nation. At Monticello, he could roam the universe.

Jefferson fixed his house, like his mind, at the height from which he could view the panorama of life. It crowns a steep, densely wooded, wind-whipped mountain. Like its owner, it heeds the classics and recites them in their original—Doric, Ionic, and Corinthian columns, friezes from Greek and Roman temples in every room, a large, stately dome, geometrical borders, the dense logic of intersecting planes, the chaste clarity of simple infinite lines. Classical thinkers were "the first to give examples of what man should be," he observed, and Monticello salutes them at every turn.

Jefferson loved columns and capitals, but he loved them pressing out against the wild profusion of nature, which could be cleared, fenced, planted briefly, but was uncontainable. So, like their owner, long terraces surge with light at daybreak and reflect on the rich, green vernacular of Virginia. The gardens skip from carefully clumped trees and flowers to open lawns, then wild groves, passing from the general to the particular, from robustly savage to resolutely calm. Because water was scarce, he built terraced gardens with cisterns and raised hundreds of varieties of fruits and vegetables. His "pet" trees, rarities that had no roots in Virginia but came from exotic lands, he planted in a special grove where he took visitors to find cool relief. "Under the constant, beaming, almost vertical summer sun," he wrote, "shade is our elysium." As it is for lions, cheetahs, crocodiles, or wild horses. Because we've now created so much shade in our lives, we take it for granted, but it can easily become priceless as rain. On stultifying days, the shade of a large tree served as the family's drawing room, office, kitchen, or nursery.

Mainly a vegetarian, he considered meat merely "a condiment," so his vegetable garden was both a valuable laboratory and an essential larder. Thus he carefully noted the planting of asparagus, radishes, corn, and a remarkable variety of exotic greens. He grew such "new" vegetables as tomatoes, cauliflower, and eggplant, acquiring squash and broccoli (from Italy), peppers (from Mexico), and figs (from France). Cabbages and beans abounded, of course. He raised sesame for his salad oil, grapes for his wine. And he adored peas. Fifteen types of English pea grew in his garden, and he noted their harvest in his garden book with much excitement. Each year, he and his neighbors competed for the thrill of producing the first peas of spring. The winner would invite the others to dine on the season's new peas.

It's difficult to locate Jefferson's heart, but here are three clues from Monticello. As you approach the front door, with its fan glass windows and classical columns, a clock face stares out with the urgency of time. Just above it, a weather vane swivels like fate in the light mountain breeze; its pointer is a heart. On the porch roof, a wind compass, which Jefferson designed, fed his meteorological curiosity while he stayed dry indoors; its pointer is also a heart. In the gardens, the broad, heart-shaped leaves of the Chinese catalpa, one of his pet trees, dance furiously in the wind, as if nothing so slender as the present could contain them. Jefferson died on the Fourth of July, around noon, on the fiftieth anniversary of the Declaration of Independence. Later that day, John Adams, hearing that Jefferson had died, said on his own deathbed, "Thomas Jefferson still survives."

Heaven knows, he survives in my garden this afternoon. Not just in my daydreams, but in the many trees and flowers I relish, which he hybridized, combining native varieties with exotics.

Trading seeds and cuttings and botanical observations with other gardeners throughout the world, he put the relative importance of the presidency in perspective when he argued: "The greatest service which can be rendered any country is to add a useful plant to its culture."

How I wish I could share my spring bulbs with him. I know their frills and colors and scents would delight him. He'd respond to the beds of pedigree and rustic roses, most of them very different from the thirty single-bloomers he planted "round the clumps of lilacs in front of the house." He'd enjoy the acre of trees, too, although it's an unruly sprawl of oak and hickory, and Jefferson was a fetishist about trees. The father of American forestry, he would no doubt be an environmental activist today, the sort of person denounced by some as a "tree-hugger." Once, while dining with the president, Jefferson insisted: "I wish I was a despot that I might save the noble, beautiful trees that are daily falling sacrifice to the cupidity of their owners, or the necessity of the poor. . . . The unncessary felling of a tree, perhaps the growth of centuries, seems to me a crime little short of murder." Among his tree plantings, he designed an elaborate eighteen-acre grove to be an ornamental forest of pruned and thinned trees, chosen for their varied textures and foliage, completed by attractive thickets and stumps placed "where they might be picturesque." Perhaps we would trade visits, letters, and seedlings. Two months before his death, at the age of eighty-three, he was still designing treescapes—an arboretum for the University of Virginia—though he knew he would never see its tall spires. "Too old to plant trees for my own gratification I shall do it for posterity," he wrote while in retirement at Monticello, adding with a poignancy timeless and endearing: "Though an old man, I am a young gardener."

8

Every morning I wake to a surprise in the garden. Overnight, new visual gems have opened, sometimes in hordes. And if, heaven forbid, I go away overnight, I return to a garden I barely recognize. While I was away a day and a half, hundreds of purple ajuga flowers sprang up in the shade of the apple trees. The Jacob's ladders—bushy yellow-and-green-striped plants—grew sprouted tentacles topped with delicate blue flowers. The first roses of the season have unfurled—two pink spicy-smelling Paris de Yves Saint Laurent. "*Mes roses,*" I say as I cut them, echoing Cocteau's rose-adoring, sensitive beast in the film *La Belle et la Bête.* Something dangling in the redbud catches my eye—wisteria! I can't believe it. It hasn't bloomed for years. Last winter I cursed it as a Medusa, gave it a severe haircut and amputated trailing limbs that had crossed the stone patio and begun gripping the entire house—and it repays me with flowers. Grapelike clusters of mottled purple-and-lavender flowers cascade from the branches. Wisterias splurge on travel. If they can't travel, they make do with blooming. It took me years to notice that simple truth, probably because I was fascinated by the gothic effect of the vines, which were creepy in both senses of the word. Every time I turned around, they seemed to have grown longer and stronger and have more of the yard in hand. I half-expected them to make a lunge for my chair. Anyway, much as I enjoyed the amusement of its macabre growth, I wanted wisteria blooms more, so last year I amputated its longer limbs.

Feathery tulips are still going strong, and I bring in several dozen. Near the Japanese lilac, I scout two clumps of hot-pink tulips and my gaze snags on what might be a slug high up in the cup of one. Slugs do climb. Peering closer, I discover to my amazement a tiny green frog sleeping inside one of the large pink petals. My presence doesn't wake it. It looks healthy and alive, just asleep. Watching it for long moments, I find myself grinning. This is a seriously adorable sight. I didn't know frogs napped between tulip petals.

Nearby, a crop of white lily-of-the-valley rings its miniature bells. Richly scented with a perfume somewhere between musk and apples, lily-of-the-valley is an heirloom flower—gardeners often dig some up and give them as gifts. I received the progenitors of this crop from my parents twenty-five years ago, when they lived in Pennsylvania. Lily-of-the-valley is comfortable here, in the moist shade beneath an apple tree. But most shade plants are, since the fallen apples have rotted over many decades and enriched the clay soil to the point of stupefaction. They get a little morning sun, and not so much shade that they grow tall and floppy. Each year they return like circus ladies—perky, talced, and heavily perfumed.

Finally, I follow the primrose path (planted with Japanese primroses—large tiered flowers atop a tall stem), behind the apple trees and through the gate to the side yard. Two bushes outside the bay window are thick with white, golf-ball-size pom-poms. When the garden starts thickening up you have to look closely to spot new arrivals. Sometimes it's hard to see the garden for the green. Flowers come and go so quickly in late spring that it's like watching a parade. Where are the magnolia blossoms that marched triumphantly? Grounded by pelting rain. Many of their shriveled corpses still cling among green leaves.

A thunderstorm moves in; the sky turns plum-black and the air feels steely. Lightning crazes the dark lacquer of the sky, and then rain falls in thick drapes. Abruptly, it stops. Baptismal sprinkles follow, then hail. Somewhere a tornado is prowling the county. Another downpour. The garden needs rain, but it also needs time to swallow. We've had many heavy rains this season (in fact, I haven't had to water), and the next day I'd find the flowers parched. A wild storm's pulse-revving intensity feels exciting and seems nourishing, but all it does is cause runoff and just worsen the land's thirst. A slow, steady seeping saturating rain is something poets and gardeners alike need, which is why nineteenth-century poet Gerard Manley Hopkins once prayed: "Oh, lord of life, send my roots rain."

In the kitchen I find a two-spotted ladybug gamboling across the countertop, flexing her wings. A glossy red carapace opens like a hangar, wings unfurl, then she bails out backward in a short jump. Landing upside down, she teeters on her shell, kicking her six feet in a tantrum. There's nothing pedestrian about her. But which leg moves first? Which side? Two white dots look like dribbles of mold. She cleans her feelers. Where could she be off to on this wide tundra? Where is her out, her leafline? She parachutes a body length or two. Her shell shines nailpolish-red—a tint called something like Ripe Tomato. Dragging her bent back legs, as she was designed to, she spins round to quest, fixes me solidly with a naked doughy eye, and yanks in her head and six legs, trying to con me. Picking her up carefully, I carry her outside and set her free to whatever predator she knows, something smaller than I and pitiless. I'm not sure who decided ladybugs were lucky, but they're welcome guests in a garden, since they eat such pests as aphids and scale. Some people import pails of ladybugs, turn them loose, and are surprised to find they all fly away. Although I've never added foreign ladybugs to my yard, a chap who sells helpful insects tells me that it's wise to buy them in their juvenile state so that they can grow up in the yard and sense it's their home. True for ladybugs, true for golden lion tamarins (adult immigrants have a tougher time adapting to a new rain forest than their young do), true for people, because the young absorb sensations that help them map their world.

There's a cricket head lying on the flagstone, probably left by a toad. Every spring, male crickets risk all for sex. As happens among some birds and bats, whales and humans, females are attracted to males who serenade them well. Male crickets chirp songs for up to three hours, and they chirp faster when the temperature is rising. One way to figure out the temperature is to count the number of chirps in fifteen seconds and add 37. Unfortunately, a robust song also attracts hungry toads. What is a cricket to do? A researcher at the University of California, Davis, discovered that "those reckless, crooning dreamboats have a cautious side," and stay hidden as much as possible, whereas the males with short chirps go exploring and are generally bolder. At the moment, cricket song is a solid wall of noise like the whine of machinery. To us, I mean. Crickets hear in ultrasound, beyond the range of our ears. Hidden in the noise are irresistible arias of loneliness and desire, sung by males with varying degrees of finesse.

9

I write in a study whose walls are the yellow of spring light in the forest. High ceilings, art deco lights, and stained glass make it look a little like a Frank Lloyd Wright sanctuary. I've highlighted white wooden bookshelves and filing cabinets, as well as the skirting boards, with a dark blue-purple. Oriental rugs, floral fabrics, potted orchids, and a swan motif add to the garden feel. My desk looks out onto the backyard through a large window shaded by a Japanese maple whose leaves vary from brownish red to olive green. But my favorite spot in the study is the bay window, which floats next to the magnolia tree and among flower beds. I've had fun making sure certain flowers are ideally visible from that perch. I sometimes recline there on yellow and purple pillows, for hours, drinking tea and watching the birds and flowers. It's my favorite place to meditate, read garden books, or nap.

Gray woolly clouds have been dumping rain for nearly a week. At one o'clock I notice the sky brightening a little, and in that lull between prowling storms, a house wren perches in the magnolia tree and serenades the yard with an exuberant trill. Its liquid song is one of the loveliest spring sounds. The description offered by nearby Sapsucker Woods captures its charm: "A high-pitched, rippling laugh that tumbles downward in pitch in 3 stages." Sometimes it also includes a fast calypso of chitters followed by musical gurgles and trills. A tiny cinnamon bird with a perky tail, the wren hops on a branch near the gourd birdhouse, eyes it carefully, then flits away. Soon it flies slowly past the gourd, settles in the tree and warbles for a few minutes. As the sun comes out, the wren flies straight into the birdhouse, inserting itself smoothly through the wren-size hole. Then it pokes its head out and jumps onto the gourd-neck perch. I hope he sets up house there, because I've got the gourd angled at eye level about ten feet away. Wrens have a very small territory; they're like city dwellers who rarely leave their neighborhood. I'd love to watch a family of wrens to-ing and fro-ing. But there are three other empty birdhouses in the front yard for wrens to choose among, each offering different advantages and views. A larger gourd hangs high against the trunk of the sycamore. A log cabin birdhouse, full of last year's twigs, and a square federalist birdhouse that I spring-cleaned, both hang from lower branches of the same large sycamore. Chickadees have already nested inside a hollow in the tree trunk. The opening to their world looks small, but sycamores can have100 feet or more of hollow space inside, a toasty campsite for birds. During the afternoon, the wren visits each empty birdhouse a few times. It's a little like a human looking for a new house—drawn to this one's location, but that one's porch. Wrens are flexible about their homes, as *The Audubon Society Field Guide to North American Birds* points out:

> This wren often nests in odd places such as mailboxes, flower pots, and even the pockets of coats on clotheslines. . . . If wrens return in the spring to find an old nest still in place, they usually remove it stick by stick, then proceed to rebuild, often using the material they've just discarded.

When the sky clears, I begin the day's chores. It feels savage to chop off healthy plants just when they're surging up thick and green. The euphemism we use is "cutting back," when what we mean is *cut them off at the knees!* Of course plants don't have knees, any more than bees do, but it's the bee's knees to have cushions of colorful flowers in fall. So now is the time to tame asters, turtleheads, boltonias, false dragonheads, and other fall-blooming plants. Cutting them down to size in May will force them to regrow bushy rather than tall. Few things are rowdier than waves of tipsy, human-sized turtleheads falling all over themselves and shorter plants (which includes nearly everything else). Monster plants are virtually unrestrainable. By the time you've strutted and fenced and tethered and staked and hooped each of the spastic church-steeple flowers, the result looks more like a synchrotron than a garden. Not that I have anything against synchrotrons. I once passed a very pleasant morning bicycling through a synchrotron's tunnels with a physicist friend, Persis, whom you'll meet later. I often see her gardening next door, and I suppose I could join her, but I always resist. I'm not sure what garden etiquette is, exactly, but I believe one rule should be "Don't kibbitz." Or, if you prefer, "Mind your own garden." People garden for different reasons, and some enjoy learning on their own, making fun mistakes, mucking around in a pile of dirt without having to explain why to anyone. If my neighbor wants advice, she'll ask.

Another rule should be "It's impolite to deadhead someone else's garden." This happens more often than you might imagine. "Deadheading" means removing blossoms after they've bloomed. One does this to confuse the plant into thinking it hasn't finished the flowering stage of its cycle yet. Otherwise it would shift its energy into deepening roots, or gathering chlorophyll through the open palms of new leaves. Controlling plants in this way is a little like keeping pets: we require them to adopt our schedule, eat and sleep when we ordain, and even go to the toilet where and when we find convenient. We impose our lifestyle on our pets, but also on our plants. I can't resist fully budded rosebushes early in May, even though I know they were forced into bloom in a greenhouse. Early in the season, I don't care if my plants are shallow, as long as they're showy. Later I'll urge them to fortify their roots, perhaps, or guzzle sunlight. But it's not all one-sided. I'll protect them from insect and drought, itchy deer or horses who want to use them as scratching posts, and of course the spine-cracking blasts of winter. I might even help them procreate. But, mainly, I'll fuss over them.

Gardeners tend to fidget with a yard, deadheading automatically, without thinking sometimes, and when we visit another's garden we continue fidgeting, automatically deadheading or weeding the way sheep idly graze. Caught in the act, one might laugh it off as "being helpful." But we all know the fanatical truth. Faced with an undeadheaded rosebush, I'd have to put my hands in my pockets, and even then they'd probably be twisting and plucking. One can't leave a spent bloom or the plant will seem to be the rough equivalent of constipated.

It's funny what things we identify as enemies. Dandelions, for instance. Mushy daylilies or slightly putrid rose petals are easy to categorize as loathsome, but dandelions? Dandy lions with yellow manes? Puff balls that predict love's truth? Let me praise dandelions. They carpet the fields with a yellow so loud it croons in the sun. They're streetwise and hardy, thriving in sidewalk cracks as easily as they do in a topiary garden. Their greens taste good, and they make a light, summery wine. Heaven knows, they're persistent. In late fall, when all the color has drained out of the grass, the brilliant flowers have vanished like a mass hallucination, and even the trees have forgotten how to green, one still sees dandelions gamely blooming. Their long, thick taproots grab deep and send off fine rootlets, divining water other shallow plants miss. They're fun to play with when they go to seed, because the lightest shake or breath launches their parachutes. They're quaintly named. The French thought their serrated leaves looked like the teeth of a ferocious lion, or *dents de lion,* which the English misheard as dandelion.

And yet, only poison ivy is more despised. Why are dandelions regarded as the pinwheels of Satan? Perhaps they're too short to qualify as model flowers. The same shape and yellow on a lanky stem—coreopsis, for example—we find classy. They multiply fast because they gush with nectar, and manufacture loads of pollen, so all sorts of insects visit them and spread their pollen. But they're also self-contained, and if no insects were to visit they could still produce seeds without pollen. One way or the other, they quickly dominate a yard. I suppose they're too ordinary to prize. Common as ants, they're easy to overlook. Some people may feel they ruin the solid green method of the lawn. I've watched many homeowners patrol their lawns like serial killers, flashing a pair of scissors with which they methodically decapitate every dandelion, and still they search for more trophies. I'm tempted to stop at those yards, perhaps holding a placard that says "Dandelions are flowers, too!" But that would contravene the previously mentioned first rule of garden etiquette: Mind your own garden.

And my neighbors may already think I'm a little strange. After all, I did once capture all the squirrels in my backyard, give them necklaces (pink beads for the females, blue for the males; both got ear tags), hand-feed them five kinds of nuts, and watch them enjoy a circus of squirrel feeders while I studied their habits and personalities. The squirrel population soared. Fifty or so would come running at dawn when I summoned them by singing out "Squirrels, squir-rels," in my best Snow White voice. Neighbors reported seeing squirrels standing up at their windows dressed like tiny Masai warriors. They understood that I was conducting a *National Geographic* gray squirrel study, and approved, but I suspect it hinted at a tad of eccentricity in the household. What must have clinched that suspicion was the day that Paul was walking sleepily down to the mailbox at the end of a driveway flanked by a small grove of quaking aspens. Hitching up his boxer shorts with one hand because the elastic waistband had loosened, he let go to open the mailbox, and his shorts fell to his ankles. Unfortunately, at that very moment a neighbor lady appeared, walking her dog. For a frantic second they stared at each other. Then she said, "What lovely quaking aspens."

"Yes, they are, aren't they?" Paul replied, hoisting up his drawers, and returned to the house. The quaking aspen event made the rounds of the neighborhood before the dew had dried on the crocuses.

I don't know what they think when they see me playing with fall leaves or standing with my nose to thick beams of white oak, inhaling the clean green fragrance. I think the neighbors mainly appreciate this oasis of flowers at the end of the cul-de-sac. I often see them strolling by to admire what's in bloom, and some say it's the destination for their early-morning walk. There's a grandmother and a little girl who visit regularly, and a young married couple holding hands. They're always discreet, and it's fun watching them from the house, seeing their faces, their casual gestures, the flowers that attract them on a particular day. It's a relaxed, private moment for them, communing with flowers on the ledge of the morning. They never step over the property line or snatch any samples, although some may well be seasoned garden thieves.

The garden thief always carries a plastic bag and scissors discreetly hidden in purse or pocket. Although he or she knows better than to steal from shops, a pod stuffed with seeds or a young fertile branch offer irresistible plunder. Gardeners get a special thrill from this petty larceny. I've never heard of anyone actually digging up bulbs from public gardens or gathering a bouquet, but it wouldn't surprise me. Another rule of garden etiquette is "Never steal samples from a friend's garden." Compliment a gardener enough on a plant and you're bound to receive a cutting, or if it's abundant, a whole plant. Gardeners don't find digging up flowers and giving them to friends a loss. More akin to a potlatch, it's a form of display. "Look what glorious gardens, I have!" such gifts proclaim. "I'm overflowing with flowers, they're multiplying so fast I have to *give* them away."

Of course, some plants multiply without any encouragement and will travel like a green brushfire if you let them. Bamboo. Morning glories. Peppermint. They quickly pass abundance and become infestation. Peppermint makes a delightful ground cover, though, and I try to plant it where I know I'll be walking, so that I'll release its oils as I pass and my shoes will be fragrant. I have two kinds of peppermint—one straight, one curly. The curly looks like short kale with an attitude, and its scent lingers longer in the back of the throat. Probably because its volatile oil contains menthol, Chinese medicine labels mint a "cooling herb," one that lifts the spirit and clarifies the mind. It's hard to think sad thoughts when you're inhaling mint. Hot and sweaty from the garden, I like to gather a fist of mint, take it indoors, crush the leaves roughly to free the oil, and then stuff them into a large tea ball, which I toss into a bathtub for a mint soak. Mint gets its name from the Greek nymph who tried to seduce Pluto, god of the underworld. When his jealous wife, Persephone, found out, she turned the hussy into a mint plant. The aroma still seduces, so much that we add it to toothpaste, mint juleps, candy, and thousands of foods and products. People started drinking mint juleps in colonial days, because the water wasn't safe. A common breakfast drink, the julep was made from whiskey, brandy, or rum, tamed by a little sugar and diluted with a little water. Mints are among the easiest plants to identify because they have square stems. Feel them carefully and you'll see. Many of the herbs we love are in the mint family: rosemary, sage, thyme, basil, oregano, lavender, hyssop, horehound, and catnip.

No month is as reassuring as May. March blows hot and cold, April feels tentative, but by the middle of May spring truly exists again, nature will renew itself, and warmth and hatchlings will follow. The natural backdrop of our lives will continue in a predictable way, regardless of our own personal maelstrom. You can trust the thickening green of the woods, the patrol of the bees, the hummingbirds up to their cheeks in fuchsia blossoms, the blue nectar-rich violets with heart-shaped leaves, the early sunrise and dawn chorus. There's no going back. The earth has changed seasons once again, as it always does, despite our tendency to doubt in late winter. Then lilacs sweeten the air on every street, some old bushes spilling over the tops of houses. I wish they were still here when the roses and yellow rockets bloom, wish the whole hallelujah chorus would coexist. After the rainiest spring on record, the once-bare trees have unfurled a million leaves to become food factories, and the whole world suddenly seems to be dyed green.

10

As the sky whitens from night purple to forget-me-not blue, the lonely whistles of a single cardinal give way to a pandemonium of birdsong. It sounds like all the members of an opera chorus rehearsing different tunes at the same time, oblivious to one another. But, of course, I'm missing its subtleties. Birds can distinguish more notes per second than humans can. If you play back birdsongs at a slower speed, you can sometimes hear the extra notes.

Clouds of leggy forget-me-nots stand out at the fringe of the woods. Most of the tulips shed their last petals during the night rain, but kept their empty stems. It's a cold misty spring morning, and today is the annual plant sale at the local high school. By general agreement, no one risks annuals until May 20, when twenty or so local growers gather to sell their best plants. I usually buy geraniums, hand-reared and inbred as poodles, and pansies sporting trickster colors. Some genetically altered pansies no longer look as pensive as they did when they were named by the French (*pensée*). The pansy has other names, too: heart's-ease, three faces under a hood, tickle-my-fancy; and in Shakespeare's day it was called love-in-idleness, which meant love in vain, and supposedly contained a violent love potion.

There will be lectures and displays, too. This year I've brought my red Radio Flyer wagon. I remember how exhausting it was last year to be hauling heavy potted plants around the grounds. But

I'm not the only one with a red wagon, and more enterprising still are the dozen people who have brought baby carriages and strollers—minus the babies—in which they transport everything from trellis-climbing herbs to potted peonies. This invites jokes—"Cute baby, takes after its father," etc.—and adds a certain zaniness to the general stir. One grower has raised gorgeous asters, already blooming and heavily budded. Some of them are even striped. Irresistible. I buy eight to plant among the ornamental grasses lining the driveway. At another kiosk I spot large-leafed coleus with a splotching of blue-red and cerise and buy four of those for the courtyard bed which will soon include many impatiens (also known as "busy Lizzies"). If only the rains would stop. News from England says they're having the wettest spring in a hundred years. We are, too. I'm tempted to build a flower bed in the shape of Noah's ark and fill it with topiary animals.

I'd call this a gentle rain, but no rain is gentle. The shape of rain changes as it falls. Raindrops are never round for long; they're high-speed shape-shifters. Pulsating wildly 300 times a second, they become domed, flat-bottomed, elongated, egg-shaped, swollen, skinny, flat, pill-like, tall. Watching a large raindrop fall in slow motion, I see the well-rounded peace of a tiny world, not furious crack-ups and mutations. It reminds me of the gradualness of growth, biological or personal. People appear to be whole, and yet they are countless small processes holding one another in equilibrium. Wishing to move beyond some outworn behavior, one might lament, "I'm not there yet," without realizing that there isn't a *there* there, since growth happens in minute increments. A time may come when one says, apropos of an awkward incident, "Oh, I feel like I'm reacting differently than I used to," but one never knows how one got there. A thousand tiny oscillations were happening in every round moment. It's like the shape of rain, which is constantly changing, yet staying recognizably the same, though at times it may even have assumed opposite forms.

Back home I resume my watch in the bay window, binoculars at my side. I don't have long to wait for drama. On a branch in the magnolia tree, the male wren begins singing a harsher, squeakier song while vibrating his wings. Must be he's wooing a female. After a few moments of scouting the tree branches, I find her, wearing the same plumage he is, sitting a few yards away. The tuneful wren has found a mate, and both sit like ornaments in the magnolia tree. Then he leads her to each of the birdhouses in turn, flying with the slow heavy wing beats that signal courtship. A little later, she flies to the federalist birdhouse and stays for a long while. I guess she's chosen that one to nest in, perhaps because it's clean and farther from the windows. While the female does indoor chores (arranging twigs, grasses, feathers, spider egg cases, and such), the male sits on his song post and warbles a territorial solo. It seems like an awfully loud song for such a little bird, but birds supplement their lungs with air sacs, which the chest muscles open and close like balloons. In time, he'll end up cleaning out new digs for a second brood, but for the moment he's all music. Sometimes he varies it with a quiet nest song addressed to his mate or a burst of chatter when danger threatens. Like the Aborigines, house wrens have songlines that map their landscape. Perching in two or three spots, they claim the intervening space with song, and heaven help intruders. I've watched a bad-tempered wren hazing a rabbit across the lawn by dive-bombing straight at it, beak ready to impale. There's no way the rabbit can climb after the chicks, but wrens are pint-sized scolds and highly territorial. Neighboring males whose songs overlap take turns singing. Soon the female will lay five or six eggs, and with any luck I'll hear the hatchlings peeping, see the parents bringing food or taking out the fecal-sac diapers.

Then something flutters at the window. The first hummingbird of the season! A ruby-throated beauty, it's a whir of iridescence, trying to feed from anything red it can find—the dial of the window thermometer, a geranium petal stuck by rain to the glass, and finally, the red-topped window feeder I've filled with sugar water. A little-known secret about hummingbirds is that some die in their sleep. Swiveling at high speed among thousands of blossoms, hummingbirds devour vast stores of energy to fuel their flights. Frantic for food, they must eat every fifteen minutes. Courting and dueling, feasting and darting, hummingbirds burn life at a fever pitch, with hearts drumrolling at 500 beats per minute. That leaves them dangerously exhausted, and at day's end they enter a metabolic limbo, when breathing grows shallow and the heart slows. What a struggle to restart their racing engines at dawn. Waking up becomes a death-defying act, and so some die in their sleep. But most, like the red-bibbed gent at my window, survive. It's a quiet visitor, since hummingbirds mainly communicate through dance, coats flashing in the sun. They do sing a scratchy little song, which isn't very tuneful, but for the most part they dance, and they're extremely agile. Not only can they hover, they can fly backward and upside down. Although they can use their frail feet for perching, they don't walk well and prefer to fly, even if it's to a spot only an inch away. The male's shimmer and swooping dance draws a female to mate, but the trembling colors aren't true pigments, they're prisms. Each feather contains tiny air bubbles separated by dark spaces. Light bounces off the air bubbles at different angles, and that makes the colors wobble.

Many things catch light prismatically—fish scales, the mother-of-pearl inside a limpet shell, oil on a slippery road, a dragonfly's wings, opals, soap bubbles, peacock feathers, metal that's lightly tarnished, the wing cases of beetles, spiders' webs smeared with early-morning dew. But perhaps the best known is water vapor. At a misty waterfall, or in the surf, or when it's raining and the sun is shining, light hits the prismlike drops of water and is split into what we call a rainbow. Today's hummer doesn't flash its neck feathers. Low light and rain clouds dull the display. But it whirs like a windmill.

Sleep seems dangerous to us, too, sometimes. Waking from a deep sleep feels like climbing out of a well. The most dangerous time of day for humans is also at dawn, when most strokes and heart attacks occur, because the body must rouse many different processes, and that can make blood pressure soar. Reptiles are only capable of light sleep. Among birds, one starts to see the beginning of brief deeper sleep.The hunting species (humans, cats, dogs) enjoy more deep sleep than the hunted (rabbits, deer). So I think deep sleep must have been a rather late development in evolution. Still, we can nap like a rabbit or a swan.

We may not near death while we sleep, as some hummingbirds do, but we do live on the edge, in a very narrow range of temperatures. Humans can survive only at a maximum of about 140 degrees Fahrenheit, a temperature that is often reached in the southwestern desert or in North Africa. In high humidity, 115 degrees Fahrenheit can kill us, and summer temperatures of 115 degrees Fahrenheit are common in the Midwest and along the East Coast. So there's only a 20-degree margin of safety in most places, even less in some, and that's precious little if the planet continues warming.

As spring becomes summer, the shape of rain gives way to sun and shade. Solid green walls have replaced the see-through forest. The thirteenth-century Persian poet Rumi once mused: "This outward spring and garden are a reflection of the inward garden." Consider the inner garden, that secret glade filled with daydreams, feelings, and memories. Sometimes they can be made physical, a goal of metaphysical gardening. Otherwise they remain mere reflections in a gazing ball. Beneath two tall spreading apple trees, I planted a primrose path that is quickly becoming my favorite shade garden within the greater garden. A garden always includes many smaller gardens. Indeed, no garden exists as a single thing. By its nature, it is plural, just as each person is a symposium of cells, or an arch is a strength made from many weaknesses. The apple garden began simply, but now includes several layers of flowers and shrubs, from rhododendrons and fiddlehead ferns in deep shadow to tulips, peonies, and daylilies in raw sun. And, of course, there are the primroses, a special variety of Japanese three-tiered primroses in which pink, apricot, or white flowers appear in rings around a tall central stem. Clematis climbs the fence, and broad-leafed hostas lend a prehistoric feel. The bark-mulch pathway invites meandering and quickly draws you to the heart of the apple tree garden, where a bench waits, surrounded by the narcotic of lilies. Do gardens provide escape? They offer sanctuary, the way cupped hands shield a flame from the weather. Into those green hands, a gardener presses her soul. Some days in late spring, while I'm napping in the apple garden, a secular prayer drifts through the thin membranes of consciousness: "Our garden, which art in flower . . ."

11

Through dense fog, the garden emerges as islands of clarity. A cat appears leg by leg. At what stage does the brain say, "Oh, yes, a cat"? My first chore of the day is to change the hummingbird's water, as I do every other day, because I want its food to be wholesome and delicious. For all I know, stale sugar water may smell as bad to it as old fish does to us. Something must warn it not to eat what will harm it. For us that something is what we call *disgust*. It's a reflex that works well, even if at times the brain uses it for funnier purposes, as when a preteen sees her parents smooching and says: "Oh, gross!"

This long drink of nectar must seem a windfall to the hummingbird, which can belly up to it and quaff for as long as he wishes. Most flowers ration their nectar, producing only small samples throughout the day. Otherwise, they might monopolize a visitor, which they need to move along to as many flowers as possible, taking and leaving a little pollen at each port of call. The nectar is a bribe. By setting the bribe rubbably close to their reproductive organs, flowers ensure their genes will travel, and the bait costs them little, since a typical apple tree produces only one ounce of nectar each day.

Overnight, the temperate jungle of the garden has changed again, as it does each day now, because summer is waiting in the wings with stage whispers, costume changes, and bouquets. Roses, roses, everywhere! On the rose garden fence, fifty red and pink roses have opened, creating a roller coaster of color. The tea roses have begun to bloom in unison, and I'm able to bring in dozens—pink climbing Colette, a fountain of buds; orange-and-yellow Joseph's Coat; musky, orange Singin' in the Rain; red climbing Champlain; red Cardinal's Song; magenta-red Othello. I love a rose flood. I bring in two dozen and leave three dozen more on the bushes.

Meanwhile a surprise greets me when I raise the pleated white curtain at my bay window. Two wrens are considering the gourd birdhouse hanging beneath a canopy of magnolia leaves, only a few yards away from my window. Over and over they play out this drama: the male flies to the house, loudly warbles and quivers, dives into the house, emerges, flies to a near branch. The slender missus approaches tentatively, warbles quietly, enters the gourd and flies out. The male often seems to be coaxing her back to the box, warbling encouragement until she enters it. For hours, they repeat this threshold ceremony. Then at 3:30 p.m. something different happens. The male perches and warbles, as before. He's close enough that, even without binoculars, I can see his neck ripple as he sings. Then he pops into the house and flies out. The female follows suit. When the male returns he carries a piece of what looks like dry grass, the first element for nest-building. From then on, they're relay builders, arriving with bits and bobs, diving into the gourd and lingering a while, still warbling to one another. Not twelve yards away, another wren couple continues feeding their brood in the federalist birdhouse. Two nesting pairs so close to the window is more than I'd hoped for.

Lightning silently flashes, and I count out three seconds until thunder. A storm is passing over the inlet downtown. Soon pelting rains send all the yard birds undercover. When it stops, half an hour later, the hummingbird appears and takes many long drinks. How does it survive when it can't risk feeding in such heavy rain? According to weather lore, plants and animals can detect coming storms. I haven't noticed that firsthand, but I don't doubt it. Their senses may well be attuned to subtle changes in barometric pressure, ozone, or charged ions. Here, gleaned from *A Book of Weather Clues*, compiled by Diane Kaiser, is some of my favorite storm lore: Pigeons bathe before a rain. Cats dance, sneeze a lot, or wash their fur against the grain. "When a cow tries to scratch her ear, It means a shower is very near." Sheep box each other and get frisky. Rabbits sit like statues and twitch their ears. Dogs eat grass and dig holes. Ants close their anthills. Fleas become more vociferous. Pinecones close and wait for dry weather. "Dandelions, tulips, chickweed and clover all fold into themselves before a storm."

In the dappled shade under a large spreading evergreen, I find a profusion of jewelweed. Yellow, orchid-like flowers hang out their tongues, ready to lick passersby. I think of them as tongues, but I guess others have found them more gemlike, because another name for the plant is lady's earring. Each graceful stalk has two serrated green leaves and, at right angles, two softly rounded green palms. We sometimes forget that nature is a well-stocked apothecary shop. Last week I used jewelweed sap on a poison ivy rash between two fingers; after one day the blisters disappeared, and after three days the entire rash was gone. All I did was cut open the stems and smear the sap onto the rash, but some people say it's best to grab a handful of the plant—flowers, leaves, stems, and all—and smash that between your palms so that you make a sappy bolus to apply to the skin. Poison ivy is an allergen whose blisters contain histamine, not poison. Only the plant's oil can be spread to different body parts or people. But, like hives, poison ivy can travel through one's system and pop up in the dark moist places it prefers.

Mid-June is a time of peeping nestlings and chattering aspen leaves, of wild and tame strawberries, and the country smell of cut hay. The birds sing at dawn as if they were auditioning and then quiet down in the heat of the day. The house wrens feed 500 insects (usually spiders and caterpillars) to their nestlings during the afternoon. In the same time, a little woodpecker such as a flicker will eat 5,000 ants. When the Japanese beetles arrive, each starling and swallow will carry thousands of them back to their nests (and still leave enough to plague the roses). A barn swallow will feed its young 500 leafhoppers before dusk. Each little brown bat will eat about 600 insects a night. I haven't begun to exhaust the long roster of insect-eating birds, mammals, and amphibians. Yet there are insects aplenty to keep entomologists happy. If you looked at a pie chart of the life-forms on earth, you'd discover that insects make up three-fourths of the biomass, about a million different species. Each year, another 5,000 or so insect species are discovered. This is good news for the birds, of course, and children and naturalists. Also good news for gardeners, because flowers and insects evolved together around the time of the dinosaurs. They need one another, sometimes so precisely as to stagger the mind. Consider the thirty species of Central American *Coryanthes* orchids pollinated by thirty different bees that respond to thirty different orchid fragrances!

This morning, seven goslings stopped traffic. On my way to a garden center, I was driving at speed down a busy highway, behind a line of cars that suddenly hit their breaks. Cars in the opposite direction did the same. Pulling onto the shoulder, I saw what they had stopped for: two adult Canada geese were herding seven fluffy newborns across the highway, single-file, as waddle-fast as their feet could take them. One adult bird led the way and the other brought up the rear. Because the goslings didn't have feathers yet, only tiny wing buds, they couldn't fly. I recognized these goslings, which I had been visiting at Sapsucker Woods, watching them graze peacefully while two adults usually stood sentry. Many people visit them, driving slowly by the small pond, and I suppose the geese are used to bad-smelling, grumbling metal creatures with talking heads. Even humans worry about crossing such a busy highway, and then it's half a mile down the lane to the goose pond. I can't figure out what they were doing at the airport. Just out on a walkabout? A training march? As soon as they cleared the highway and started down their lane, the traffic resumed, and I bet everyone was smiling.

12

The shock hit me slowly, in steps of disbelief. Lying on the grass was the federalist birdhouse, horribly silent. The whole yard hung in a bubble out of time. No warbles came from the parent wrens. No chorus of peeps from the hungry chicks. No bird sounds at all, not even from the crows. The bird feeder looked slack-jawed. A heap of nest twigs lay on the grass. Even the log cabin birdhouse, though still swinging, sounded quiet. Something had happened during the drizzly night or at dawn, something violent enough to scare away other animals.

Rushing outside, I picked up the fallen house and peered in. I could see some nesting stuff still inside, but no chicks, no blood. No blood on the ground either. Did a squirrel or raccoon bat the birdhouse to the ground, then reach inside and grab the chicks one by one? Maybe the birds had simply fledged and in all the commotion their house fell? No, it was too soon for fledging. Wren chicks usually leave the nest after sixteen or seventeen days. I began hearing the peeps on June 2. Today is June 11. The chicks would have been nestlings for another week or so. And the mother? A mother typically roosts with the chicks at night. Has she been eaten, too? When I think how terrifying last night must have been for that wren family, I feel a chill of anguish. Wrens are calling in the distance, but not nearby. Who could blame them? The other pair of wrens, the ones nesting in the magnolia tree, seem to have vanished also. I inspect the "squirrel-proof" bird feeder, completely empty, its jaw frozen open like a corpse's. Where is the second wren family? Did they leave the area because it was unsafe? Were there too many marauders, too many chick-eating thugs? What on earth happened last night?

How quickly intimacy with wild animals can grow. One becomes fond of them, sensitive to their destiny. It's easier if you don't care, don't get attached, but then you miss out on so much, and so do the animals. I always care, always suffer, and, in time, forget enough to care again for another. Not completely. I will never completely forget the penguin chick I saw pecked to death by the skua on South Georgia, an island near the Antarctic. Never forget the mother of a friend, whom I visited often and grew to love during her days in the local hospice. It's hard to love what will come to pass, be it wren or human, if you think too long on the price, which is suffering at the end. But I feel it dishonors them not to try.

I only knew the chicks by their hunger, and how bright and hearty they sounded. Now they are less than an inkling. Was it the gray squirrels? The red squirrels? Either could have been the culprit. One is not supposed to take sides, but I feel the hard punch in the stomach that is loss. All the work the wrens put into raising their young—then to helplessly watch them mutilated and devoured! What horror. I know this is the way of nature, kill and be killed, and that the squirrels and raccoons have young to feed. Such events might be bearable were it not for love, tenderness, compassion—emotions that evolved so animals would fight for survival, their own and their loved ones'. Life feels so full and continuous each day, and then without any warning, despite all the relationships, appointments, investments of time and emotion, it can vanish, leaving only a lull where a life was. "Nothing that is can pause or stay," Henry Wadsworth Longfellow reminds us in his poem "Keramos," which continues:

The moon will wax, the moon will wane,
The mist and cloud will turn to rain,
The rain to mist and cloud again,
Tomorrow be today.

In the long run, all succumb to chemistry, some sooner than we expect. There are no reprieves, only stays, and confronted with mortality the mind can only try to adjust. Of all minds, though, one that's mystical but believes in no god finds itself in the bizarre predicament of seeking not some answer or key or justification of God's ways to man, but spiritual and nervous relief: sometimes through art, in the fabrication of amulets, retaliatory enigmas, memory aides, or just elegant and distractingly beautful images. "I believe in God," Frank Lloyd Wright said, "only I spell it Nature." Gardeners spend much of their time kneeling in postures of prayer.

All morning I brood about the dead chicks. Then, after lunch, I decide to clean out the fallen birdhouse and put it up once more, this time with thick wire. I start thinking like a wren looking for safe sites. Both red and gray squirrels will eat birds' eggs and nestlings, but especially red squirrels, and they're small enough to ease in and out of the box. And there's a raccoon family living in a yard tree, too. No use dwelling on the carnage. Nature neither gives nor expects mercy. We do, imagining the suffering of others as our own, and we grow attached quickly to all creatures young and vulnerable.

Later, mourning more than I realize, I climb onto my bike for a short ride under the threatening sky. Heading north, I come to a busy intersection at precisely the moment a turning car does, and we collide. Shouldering into the huge hood, I slide beneath my bike to the ground. I've never been hit by a car before, never had a serious accident on my bike. Everything hurts: leg, shoulder, thigh, back. But I'm wearing my helmet, and nothing is broken. At home, I rub ice everywhere I'm bruised and swollen, the worst being a separated right shoulder. Although my neck doesn't hurt much, two of the vertebrae have widened in what's generally known as whiplash. All will heal. I'm lucky not to have broken anything. It was the chicks; my helplessness distracted me.

"At least they'll live in your writing," Paul says quietly. "Their lives won't have gone unrecorded."

An unrecorded life—what most plants, animals, and humans live. I would record them all if I could. In the film *The Treasure of the Sierra Madre,* bandits steal burros laden with animal hides, not realizing the burros are also carrying bags of gold dust. Unrefined gold dust doesn't shine. Assuming the sacks simply hold sand to make the cargo look bulkier, the bandits slit the treasure sacks open and toss them away. The viewer knows how many people have suffered or died for those sacks. But the real emotional punch comes at the end, when the north wind picks up and begins scouring the sky with gold dust, blowing it back to the mountain where it was mined. The traditional burial service also says, "Dust to dust . . ."

The next morning I wake thinking of a phrase that Evelyn Waugh used as a title for one of his novels: a handful of dust. In my dream I saw a small amount of dust—the ashes of what once was a human being living a complexly energetic and personal life—reduced to dust sifting through another's hand, or mingling with the blowing dirt on a country road. One's sense of one's self as the center of one's life, all life, dominates and we forget that, despite the striving, anguish, limitless sensations, and desires, we will become the earth itself. A life feels so large and sprawly, so magnetic—it attracts people and objects to it—and when all is said and done, despite the flares of helplessness or angst, it feels sufficiently controlled: it is impossible to imagine ourselves reduced, anonymous, disconnected.

Sitting beside a dry bed in the garden, I gather a handful of dust and let it trickle from my palm. A heavy, horizonless sadness fills me. *Perishable.* We should arrive with the word stamped on our outer wrappers. Perishable as a freshly baked loaf. Yet our problems seem so large. They seem to matter. That is, they seem solid and truculent as matter. Getting my mind around that fundamental truth is as hard as encompassing infinity, even though a vital part of gardening is learning to trust change.

13

The scales are tipping from spring to summer. The season fluctuates, it doesn't progress evenly, but grows the way teenage girls do, in fits and starts, at times still very young, at others startlingly mature. Up before dawn, I clean the hummingbird feeder and refill it with freshly boiled sweets—simple sugar water. At exactly 6 a.m., Ruby arrives, pokes his needle bill in one hole, laps for a few moments with a long W-shaped tongue, then pulls back and, lining up for the second hole, sips briefly, and finally drinks from the third hole, before he hovers backward and jets away. Only males have bright red necks. Unlike the wrens, which stay together to house-hunt, nest, and parent, and even raise a second brood, hummingbirds mate fleetingly and live a hermit's life. They don't socialize or travel in flocks. A solitary male will find a high perch and sing a scratchy warning to other birds to steer clear. I know where Ruby's perch is because I've tracked him flying home from the feeder to a certain branch on a hickory. I wonder how many chicks he may have fathered. Somewhere there will be a softer-hued female who has built a nest from lichen, spider silk, and such downy plant fibers as the fluff shed by quaking aspens when they go to seed each spring. I'd love to see her taking measurements, because a female hummingbird builds her nest to the exact size of her bottom, which can then serve as a warm lid over the young. She'll be sitting on two small eggs delicate as pearls or feeding her nestlings on insects and nectar. I haven't seen any females at the feeder this year, but that's not unusual, since females spend most of their time feeding and protecting their young, and in any case, my garden offers hummingbirds many other places to dine. Among the wildflowers I've planted orange trumpet vine, red cardinal flower, columbine in many shapes and colors, pink honeysuckle, orange-yellow jewelweed, pink turtlehead, red bee balm.

Among the other flowers: pink, red, and lavender impatiens; purple petunias; red and purple fuchsia; pink and red sweet William; multicolored zinnias; pink and white phlox; pink, blue, and purple foxgloves; red scabiosa; pink and lavender clematis; red geraniums; white hostas; pink weigela; scarlet runnerbeans; yellow dahlias; multicolored lupines; and yellow, white, lavender, and purple butterfly bushes. Plus roses in different hues. When I think about it, my garden offers hummingbirds a buffet that spans the seasons. I'm lucky they bother with my small window feeder at all! Notice that most of the flowers they prefer are red, a color as meaningful as a billboard, designed specifically to attract their attention, but they will sip from other colors. It's just that tubular red flowers often contain the most nectar, and after a while, hummingbirds discover the richest spots to dine.

Ruby would probably attack anything trying to muscle in on his food, and I do mean anything—wasp, raccoon, human, cat, butterfly, hawk, or another hummingbird. Hummers are well-armed and notoriously brave; their piercing bills could put an eye out. Yet when I encounter Ruby outside, as I sometimes do, he ignores me entirely or withdraws to a polite distance and waits for me to move. I suspect he's watched me tend the feeder from his high lookout post. He doesn't seem much bothered by the conga-line of ants invading the sweet reservoir or the streaks of pungent vaseline I've painted along window edges to discourage swarming ants. I don't mind if a few ants get through, because Ruby might find them a nutritious bonus. We think of hummingbirds as dainty little jewels, living like the Greek gods on nectar and ambrosia, but they do eat insects, too. They can open their bills wide enough to capture an ant or a gnat.

Ruby probably woke up hungry and was glad to find his first meal of the day waiting. Last thing at dusk, he comes to drink at great length, guzzling enough to sustain him through the night. I sometimes see him at this feeder during the day, but he likes to vary nectars, fuchsia and jewelweed being two favorites. The red geranium trained into a tree sometimes distracts him with bright red flowers he doesn't prefer. The red center of the thermometer confuses him. Come summer, he will lose himself in the hibiscuses, besottedly visiting them over and over. I wonder how the nectars differ in flavor to him? But I understand his having cravings for certain flowers; I feel the same way about fresh sweet cherries and blueberries. And I appreciate his high-octane hunger: a human burning that much energy would need to eat 155,000 calories a day.

I have peony envy. Okay, I admit, there are lots of gardeners' puns, a personal favorite being: "With fronds like these, who needs anemones?" And there are lewdly named plants, like golden showers or lust-in-the-dust. When it comes to digging machinery, rhyming jokes abound. There is a machine called a ditch witch, which will dig a four-inch-wide hole. Around this house, garden humor also includes a trench wench, a rock jock, a crevasse lass, and a crater mater (driven only by those educated at private schools). But I really do have peony envy. I admire the larger, more abundant peonies of my neighbors, who wisely planted their peonies in full sun and are enjoying cascades of cantelope-sized blooms. Their peonies are carefully staked, too, and some of them restrained by peony hoops, which they regularly adjust, cursing the weakness of wire.

A creeping obsession that afflicts many gardeners is the desire to triumph over the forces of chaos and disorder, battle the unruliness of growing things, cage beauty, impose a firm bridle on the natural world. This urge for order steals into a gardener's sensibility, and soon one is lamenting the frustrations of staking, mowing, uprooting, deadheading, pinching back, weeding, training, trimming, watering. Gardeners wage a secret war against nature's teeming jungle, which threatens to suffocate them. We don't want to be overpowered, digested like any carcass, and rendered down to our constituent molecules. We want to stay singular. It is a private, tidy war against death.

Are gardeners control freaks? No great deed happens without passion. Some passionate gardeners are petty tyrants. But not all. Learning to live with compromise, uncertainty, and failure is one useful goal of gardening. Sometimes I wonder what it says about us as humans that we search for perfect order in imperfect places—polling booths, sin-bins, books, churches, gardens—and can turn even the ceremonial violence of sports into the mercy of a workable peace.

Some gardeners seem unable to fully enjoy their gardens, so caught up are they in the latest skirmish with mildew or beetle. Weeding can attain the status of a holy war. My philosophy is: Forget winning, cultivate delight. In fact, I've been thinking it might be nice to turn the front yard into a summer meadow. I could move out all the annuals and perennials and plant wildflowers and weeds. There is a law about letting one's yard go native. But what is the line between a garden of pretty weeds and an abandoned field? I like weeds. They crop up in unexpected places, they adapt superbly, and they're inventive, finding ways to survive surfeit or famine. Also, they improve the soil. I've planted weeds on purpose in most beds. Common mullein (aka *Verbascum*) is a favorite, with its tall spires displaying hundreds of yellow flowers. Snap-snaps (aka white campion) is another, this one complete with tender memories. When I was small enough that my mother walked me to school, we used to pass a vacant lot where snap-snaps grew. These were wishing weeds. She showed me how to hold the flower by its five-petaled top, and strike the bladder hard against the inside of my wrist. If it made a popping sound my wish would be answered. They bloom every other year, but if you stagger plants you're covered should you suddenly need to resort to magic. Marsh marigolds provide a splurge of yellow in wet places. I've staged dame's rocket all over the garden, and from May till August it offers fragrant pink clusters of petals. Solomon's seal, Jack-in-the-pulpit, trillium, and wild orchids grow happily among the trees. Joe-Pye weed stands tall at the back of beds, and sometimes people are surprised to see it in a home garden, since it's a quintessential weed with pink flat-topped clusters of flowers and usually grows in wet thickets. Tall, white, fuzzy-tipped boneset gives height to a perennial border. Mixed with it, purple loosestrife adds spikes of magenta flowers. Its generic name means "blood from wounds," which I presume refers to the blue of its flowers or the red of its fall leaves. Loosestrife is considered a scourge in local wetlands, where it spreads like a rumor, driving out other plant species. But it's very well-behaved in my garden, keeps to itself, doesn't crowd its neighbors. However, in some states it's banned. It's a misdemeanor to plant it in Arizona, California, Colorado, Florida, Idaho, Illinois, Indiana, Iowa, Minnesota, Missouri, Montana, New Hampshire, North Dakota, Ohio, Oregon, Pennsylvania, South Dakota, Tennessee, Washington, Wisconsin, or Wyoming. In most of those states I couldn't plant Canada thistle or field bindweed, either. Technically, it's illegal to plant the opium poppy, too, but I've never heard of a gardener being prosecuted for it. Some weeds, especially, can drive out a state's native plants, ruin its ecosystem, or poison livestock, so they become banned. Would someone do jail time for planting invasives? Unlikely. But states can impose stiff financial penalties.

Fortunately, most weeds are legal, most wildflowers welcome. The wild swamp irises thriving among the often-flooded grasses that line the driveway are the emblem Louis VII chose for his shield and banner during the First Crusade in the twelfth century. In time yellow irises became known as "fleurs de Louis," which to English ears sounded like "fleur-de-lys." When Edward III conquered France in the fourteenth century, he added the fleur-de-lys to his own coat of arms. If you look at the fleur-de-lys pattern, so often associated with formality and decorum, you can see the angles of an unopened bud and the arching shoulders of the wild swamp iris.

Another weed I enjoy is thistle, which I encourage to grow tall and spiky. Wild sweetpea is probably my favorite weed, because I adore its flapping pink tongue-lolling flowers and thick climbing limbs; I've planted it on trellises, mailboxes, and fences. Jewelweed, covered in tiny orange megaphones, occupies its own corner of the garden. At first I raised it for use with poison ivy, but now I keep it lush to please the hummingbirds, which poke it for nectar.

In fact, their adoration of its nectar (and the inadvertent pollen rubdown that goes with it) is probably why it's spreading so well.

I treasure weeds, but a great many gardeners, Robert Louis Stevenson included, have waged war against them. In a letter to Sir Sydney Colvin from Samoa in 1890, Stevenson wrote: "*Nothing* is so interesting as weeding. I went crazy over the outdoor work, and had at last to confine myself to the house, or literature must have gone by the board." That he emphasized the word "Nothing," as if its absoluteness couldn't capture his obsession, gives a sense of how maniacal he became about routing green invaders.

One way I cultivate delight is to abandon myself to individual sensations, savoring them until they vanish. A garden pleases all the senses, including the kinesthetic sense of moving through space. For example, smelling a peony's blossoms until the nose quits from the sheer abundance of scent. In that moment, the universe—from the dirt below one's feet clear out to the farthest stars, and beyond that in time back to the Big Bang—all of it vanishes. Nothing exists but the citrusy smell of one peony. How long can I hold the sensation in my mind before it evaporates? I don't care. I cultivate delight.

Change the sense. The skin is singing with touch. The fingers, of course, handling the silky damp petals of the peony. But also the ridge beneath one eye, where sweat beads, ready to fall. The feet resting in plush socks. The cheek across which a single hair, lifted by the breeze, may be an insect walking. The blink to recenter a floating contact lens. The twinge as a fat pad shifts in the left knee. The drape of the cotton T-shirt around my shoulders and back. The pebble at the base of the peony—a round, rain-polished stone I pick up and gently rub like a prayer bead.

As I pull my red Radio Flyer wagon behind me, a wheel squeaks, and the sound transports me across time to the Pocono Mountains and the Susquehanna River. At Girl Scout camp, when I was nine years old, we lived in tents in the woods and ate in a log cabin dining hall. Every girl had her chores to do, which changed daily. A paper wheel would be turned each morning to reveal who was assigned to what. One chore was cleaning the kerosene lamps, so important for negotiating the woods after nightfall. Black soot was hard to remove from the glass, and I can still remember the squeaky, squeaky sound of the cleaning paper rubbing across the glass, which is also the sound of the rusty axle on my wagon. Another chore was cleaning the wooden outhouse with Pine Sol and a stiff scrub brush, and now whenever I smell pine I remember the patina of those wet wooden floors.

One of my garden chores these days is to castrate the lilies. When their flowers finish, lilies produce round ripe pods that are taut enough to crack off like walnuts, or sever with sharp scissors. I don't need the lilies to grow from seed (it would take too long), and plants can't spend the same energy twice. I'd rather they care for their tuberous roots and prepare for next year's blooms. Some of the daylilies, most notably Stella de Oro, will bloom again, and deadheading speeds the process. By the middle of June, the pageantry of spring flowers has fled, leaving husks and hulls behind. But the roses are blooming like racehorses hitting their stride, the animals are simply besotted with one another, insects rule the universe, and summer's petaled inferno is right around the bend.

Summer

Nobody sees a flower really—
it is so small it takes time—
and to see takes time,
like to have a friend takes time.

—Georgia O'Keeffe

14

Summer is a new song everyone is humming. From atop a chestnut tree, where spiked fruits hang like sputniks, comes the sound of a bottle band and the kazoo-istry of birds. On the ground, a blanket of dry leaves gives sound to each motion: falling berries, scuffling voles, a skink rising from its bog. Small fence lizards do rapid push-ups as part of their territorial display. All along the weedy roadways, grasshoppers thrash and rustle in the brush, playing mating tunes. Grasshoppers are musical instruments. They sing by scraping a row of 80 to 450 pegs on the inside of each back leg against hard ridges on their forewings. Different species have different calls, depending mainly on the arrangement of the pegs. There are alto and tenor grasshoppers, plus a band of crickets and cicadas rubbing shrill songs on their washboards. The grass has grown tall at last, and the trees offer shade for the first time in a year.

Expectant and rowdy, animals enter the green metropolis of summer through the tunnel of June, and they all have noisy errands to run. They bustle about their business of courting, warring, and dining. Only humans fret over meaning and purpose; animals have appointments to keep. Even the June bugs, clattering against the window screens—humming and buzzing, bumbling and banging—are on a mission of romance. Related to the Egyptian scarab, buried with pharaohs, they sometimes batter their way indoors by mistake, and then run around in beetle-mania like metal windup toys.

Spring meant scant food, faint light, and hardship. But summer is a realm of pure growth—the living larder of the year—full of sprouting and leafing, breeding and feasting, burgeoning and blooming, hatching and flying. Now the mallards are taxi-dancing as they ceremonially mate. Baby garter snakes lie like pencil leads in the grass. Wild strawberries ripen into tiny sirens of flavor that lure chipmunks, rabbits, deer, and humans alike out into the open to graze. A vast armada of insects sails into the rose beds, and groundhogs dig among the lilies.

Countless birds seem to be auditioning for their jobs. Large glossy crows sound as if they're gagging on lengths of flannel. Blackbirds quibble nonstop from the telephone wires, where they perch like a run of eighth notes. I sometimes try to sing their melody. Because every animal has its own vocal niche (so that lovesick frogs won't drown out the hoarse threats of a pheasant barking at a dog), summer days unfold like Charles Ives symphonies, full of the sprightly cacophony we cherish, the musical noise that reassures us nature is going on her inevitable green way and all's right with the world.

Drawn by a familiar chirp, I look out the bay window to see a cardinal couple feasting together on sunflower seeds. Scarlet red with a black mask, the male eats first, lifting a sunflower seed, rolling it to one side of his beak where a built-in can opener cracks the hull, then rolling it to the can opener on the other side to finish the job. Meanwhile, the dusky female stands nearby and shivers. Puffing up her feathers and squeaking, she looks helpless and cold, but actually she's inviting her mate to court her. Though she is acting like a hungry infant bird, she is perfectly able to feed herself, and will. This dramatic appeal is the time-honored way female cardinals (and many other birds) play house. When the young are born, the parents must feed them nonstop, so she wants a mate who recognizes the plaintive signs of infant need and knows how to respond. The male cardinal observes her display and cocks his head, as if listening, but what he's really doing is looking hard. Because most birds have poor stereo vision, they see better if they look with one eye at a time. Lifting a plump seed, he pogo-hops over to her and places it carefully in her mouth.

Meanwhile, two robins are running relays to feed their squawking brood in a nest they've placed in a yew tree near the door. The chicks are all gaping mouth and yammer, just fluff and appetite. How do the parents keep from feeding the same chick over and over? Do they somehow keep track of which ones have been fed? They don't need to. Birds have a reflex that makes them pause a while between swallowings. If a parent robin puts a worm into the mouth of a just-fed chick, the worm will sit there and not go down. Then he simply plucks it out and gives it to a different chick.

Nearby, the lavender garden is a den of thieves, as dozens of plump bees fumble the flowers. Dressed in yellow sweaters, the bees aren't stately and methodical about their work, but rather clumsy. Skidding off shuddery petals, they manage to grab a little nectar, but also get smeared in pollen as they careen out of the blossoms. Hovering for a moment, they dive headlong into the next flowers and spend the day in a feast of recovered falls. One rarely notices the uncertainty of the bee, wallowing and sliding, or how flower petals are delicately hinged so that they will appear firm, but waver and flex suddenly without actually breaking off. The purpose of the design is to unsettle the bee.

When night starts seeping through glossy dark leaves, a whippoorwill cracks the long three-stage whip of its voice, flaying the air alive. It belongs to a family of birds whose Latin name (*Caprimulgidae*) means "goatsucker," because they were often seen traveling with herds of goats and were thought to milk them dry during the night. Now we know it was the goat-sucking insects the birds hunted. But the name stuck, replacing the more common "nightjar," which is a better-fitting alias for a bird whose boomerang voice can jar the night right off its hinges.

Midsummer Eve, on June 23, falls two days after the summer solstice. Once it was said to be the witches' sabbath, when an evil spell could dishearten the coming harvest. On that night, if a maiden put certain yarrow sprigs beneath her pillow, she would dream of her future husband. On that night, bathing in fern seed could make a man invisible, and walking backward with a hazel twig between his knees would lead him to treasure. The summmer solstice is just a little sabbath with the sun. Summer officially begins with the solstice, from the Latin *solstitium,* "sun standing still." For a few days, the sun rises and sets in almost the same spot on the horizon, a prelude to the longest day of the year, and then the sun begins to crawl south through imperceptibly shorter days, toward an unimaginable winter. But for the moment it is still early, spine-tingling summer. Jasmine and pine leaden the scents of evening, spores like manna drill the sky, and pheasant eggs sneak life out of damp sod, while summer disavows any passion stronger than earth's in the sound of rain, in open field, when drizzle breaks.

15

When landscape architect Paula Horrigan arrives, we stand beneath the sycamore for a few minutes, sharing news of our lives. We do not try to explain the natural forces that guide them, the blights we fear, the nourishment we seek to continue growing. But on some level, we know the body too is a garden. Mainly water, we are barely contained estuaries of muscle, organ, and fluid. I marvel that we don't slosh with every step.

Then we get down to the pleasure of today's business and begin strolling around the yard, studying each of the beds, considering the view from certain windows, shaping invisible possibilities. Paula teaches at the unversity but also has a private practice, and I thought I'd hire her to help me design my garden, which has been evolving in fits and spurts over the years. It's tiny compared to Hospicare's grounds, where Paula and I first met. She has volunteered thousands of hours to designing Hospicare's extraordinary gardens—extraordinary because they're so poetic. In fact, I sometimes think of her as a kind of spatial poet, part of whose job is to reconcile the needs of plants and people.

What does a landscape architect do? I used to wonder. Now I know that it's not just a question of arranging trees and stones in a pleasing way. Her garden designs are full of insights about human activity. They require complex problem solving, sometimes with seemingly unrelatable objects—for example, a bulldozer and a duck who's unwilling to move her nest. As with all creativity, Paula's art requires spontaneity bound by restrictions. Her empathy for the people who will be nourished by the gardens, and her vision of our intimate relationship with nature, shine through her designs, which are often symbolic or metaphorical. Watching her at work over the years, I've discovered just how emotional landscape architecture can be. During the summer, Paula will play with designs, and then I'll most likely revise one small section of the garden at a time. This will be a long-term project. How much garden does one need? I think Nathaniel Hawthorne gauged it perfectly: "My garden . . . was of precisely the right extent," he writes in *Mosses from an Old Manse* (1846). "An hour or two of morning labor was all that it required. But I used to visit and revisit it a dozen times a day." There he would stand "in deep contemplation . . . with a love that nobody could share or conceive of, who had never taken part in the process of creation."

"Spontaneity bound by restriction," I remind Paula, since Paul is fanatical about leaving the lawn untouched. Paula understands that to Paul a lawn is a powerful image, the representation of the lush, wet countryside of his childhood home in miniature. Where he grew up, in an agricultural landscape, green was everywhere. So throughout childhood his main experience of land was open lawns and fields, crisscrossed with public footpaths. He could hike for miles through meadows, farms, or villages, encountering fields or enclosures where animals grazed. Typically, there was a way for pedestrians to get over a fence—a stile. Stiles take many forms. If you come upon a stone fence, the walls might get thicker, allowing you to climb up one side and down the other. Fences would have wooden stairs to climb up and over. These intersections between human and animals paths create wonderful moments, as each accommodates the other. Because Paul's experience of land was a great expanse of green, it's part of his primordial sense of place.

Paula's husband, Scott, a labor relations negotiator, feels similarly about his boyhood landscape. "Scott can't get northern California out of his system," Paula muses. "He can't get out of his system that change in the landscape from green to brown to yellow. It really changes dramatically! I'm usually there in the summer when it's parched and brown, but once I was there in the fall and I saw a reverse phenomenon. Here in the East we go from green to fall colors, then browns, and yellows. In California, you get the reverse—yellow, brown, to green in November. The rains start to come and suddenly all the hills turn green. You have this weird late-year phenomenon of the leaves turning yellow but the lawns turning green. It's really wild, very surreal, incredibly beautiful."

The landscape of childhood becomes an indelible part of memory. You absorb it, it absorbs you. Myself, I grew up in the Midwest, in a small town not too far from Chicago, west of Lake Michigan. I lived across from an orchard, which I loved to explore, as well as abandoned lots and vaguely tended yards. No stiles. But I'd like to have one now, which is what I tell Paula.

"I'm sure we could come up with a stile to go over the deer fence," she says, laughing, "It's so tall it would basically be a ladder! We could come up with a door that's elevated, where you could climb up, get to an upper level, and open an elevated gate."

I smile as I watch her play with the possibilities. She continues: "Then we could also have something that is characteristic of a medieval garden: a mount. They built a high point, a hill. As the garden progressed through the centuries, through the Edwardian garden, you had walls that you could climb up on to have a view from on high—you wanted to see the garden spread out before you. The French garden evolved with the tradition of the *parterre,* the ornamental carpet that actually wants to be viewed from above. So having places where you could get up high would be cool."

"Very cool," I agree, "especially if we got some beautiful boulders from the local quarry. I love boulders."

"That's where I got my rusty old cement mixer," she says brightly. Curvaceous and covered in magnificent shades of rust, it makes an eye-catching sculpture in her yard. "Reassure Paul that I, too, love lawns. Tell him that we're even going to put a sheep out there, to keep the lawn nice and clipped and fertilized!"

I decide against a hermitage complete with hired hermit, against ersatz Greek and Roman ruins in the woods, and against a ha-ha—a five-foot-deep trench around the property—although all were garden fashion in the past. But I have to give serious thought to a labyrinth, especially since I know a labyrinth builder.

In the nearby village of Ellis Hollow, there's a labyrinth maintained by the Fellowship of Light. On my way there for the first time, I wondered what a labyrinth might look like. Mythological beasts cornering people in cul-de-sacs? No, that's a maze. What I found was a circle of rock slabs arranged Stonehenge-like around a recently used campfire. Beyond that, in an open field, a complex pattern had been cut into the grass. Narrow enough for one walker, the lanes of the labyrinth curved and twisted, circling round and round each other. The goal, the round heart of the labyrinth, was visible but unreachable except by labyrinthine journey. So I began. A few yards straight, then a curving left. . . .

Soon three people showed up (including a man wearing a T-shirt that said "Tao Id Ta Tchai"), and they each paused a moment, centered their thoughts, and entered the labyrinth at varying speeds. Four of us circled slowly around the center, weaving beside or glancing past one another. It seemed like we should be on the same path, and in a sense we were, separated only by time. But that made each path different. One is always close to the heart of the labyrinth, always close to the beginning and the end, but circling around them, just out of reach. At times two or more of us intersected, walked a pace together, then drifted apart. Turning tightly, but smoothly, left and right, the body falls into a swaying rhythm. One has no choice but to focus tightly on one's path. At the center there is no change, no direction, only a uniform circle, a glade of simplicity in which motion ends. On all sides stretch the labyrinth, and beyond that formality the far more complex labyrinths of field, trees, houses, cars, colleges, town, factory, synchrotron, riding hall, volunteer fire department, and the celestial mechanics of the solar system.

On the stove in my kitchen, the filaments turn radiant red as they heat, and they are each shaped into a perfect labyrinth. Apparently, a labyrinth is the most efficient way to coil a single line into a circle. But do I want a grass labyrinth to walk in my garden? Who will maintain it? At the Fellowship of Light, labyrinth cutting (about a quarter-mile of intricate turns with a mower) is undertaken as a spiritual task. I could advertise for a Zen grass cutter, and most likely would hear from someone, given the flavor of my town. How many people do I want visiting my garden each week before it loses its rich privacy?

After Paula leaves, I feel elated. Collaborating with her—someone I can easily identify with, though we create in different genres—will be fun. Still, the idea of *designing* a garden feels a little uncomfortable. Did our ancient ancestors design their gardens? No, they lived inside them, like cells in an artery, acted upon, prey to weather and predator, feeding opportunistically. Would they have redesigned their vast garden if they could have? That's a different question. Humans have always been creative and control freaks, and indeed planting gardens is exactly what we ultimately did, millennia ago, in the agrarian revolution that made civilization possible. How decorative were those gardens? Were they simple and efficient? Did form follow function, as Frank Lloyd Wright would have wished? Or did people also plant flowers merely to delight the senses, leaving space for the luxury of ornamental gardens? Labor was precious. Would they have spent it on the extravagance of beauty? I think so. Gardens don't just please the senses, they satisfy one's need for calm, privacy, balance, and stability;

they allow one, no matter how weak or disenfranchised, to impose an order on the chaos and govern living things. As Juvenal wrote in *Satire III* in A.D. 100, "It is something, in whatever place, in whatever corner, to have become the lord and master even of one single lizard." Gardens offer opportunities for diversion or display. How large a plot of earth would one need to fulfill such yearnings? Would a square foot be enough?

Do we still yearn for our life on the savanna? E. O. Wilson, Gordon Orians, Yi-Fu Tuan, and René Dubos all suggest that this may indeed be the case. They point out that people work hard to create a savanna-like environment in such improbable sites as formal gardens, cemeteries, and suburban shopping malls, hungering for open spaces but not a barren landscape, some amount of order in the surrounding vegetation but less than geometric perfection. There were ample plants and animals on the savanna. Fish, shellfish, and interesting plants thrived in the rivers and lakes. Hills and cliffs served as fine lookout points. Wilson observed that

> Whenever people are given a free choice, they move to open tree-studded land on prominences overlooking water. This worldwide tendency is no longer dictated by the hard necessities of hunter-gatherer life. It has become largely aesthetic. . . . We who enjoy their creations without special instruction or persuasion, are responding to a deep genetic memory of mankind's optimal environment.

Chrys Gardener adds that she's noticed a distinct gender pattern among homeowners, with men preferring open areas and lawns and women preferring lots of surrounding growth. She thinks it may be a carryover from hunter-gatherer days, when men found safety in a long view across the grasslands and women and children relied on the protection of thickets.

How we translate our evolutionary urges differs from time to time, and for many reasons. We love to elaborate necessity. We grow bored easily. We feel the need to make visible to the outside world our inner lives and values. In any case, the climate and resources often change. In ancient Egypt, for example, and throughout the Mideast, where daily life felt torrid as sand, paradise was a garden promised in heaven. Wealthy landowners could plant their own private paradise on earth, and portable dioramas of enclosed gardens have been found in the pharaohs' tombs.

What qualifies as a garden? An enclosed, cultivated space? A Zen garden doesn't have to include plants, yet it's a site of spiritual growth that usually includes carefully arranged rocks and sand, exposed to changing weather. A topiary garden looks different from a wildflower garden. A moon garden is not a winter garden. A rock garden features low, creeping plants. A vegetable garden serves a different purpose than a cutting garden, even if one cuts and arranges vegetables as works of art. A sculpture garden has different goals than an herb garden. It's worth noting that there aren't any flowers mentioned in the Old Testament's Garden of Eden, only trees. But, apparently, God's idea of paradise was a garden. Rudyard Kipling made merry with this idea in his poem."The Glory of the Garden," where he chimes:

Oh, Adam was a gardener, and God who made him sees
That half a proper gardener's work is done upon his knees,
So when your work is finished, you can wash your hands
and pray
For the Glory of the Garden, that it may not pass away!

Is a maze a garden? An orchard is not thought of as a tree garden, though it is cultivated. How do we imagine a child's garden of verses? Or a beer garden? Or Madison Square Garden? Or a German zoo, known as a *tiergarten?* Or Garden City, New York? Originally, gardens were either sacred groves for various deities or vegetable gardens. As cultures changed, so did their gardens. Splendid formal histories of gardens abound, and I won't attempt to summarize them here. But we've come to want more from our gardens, more even than food, dyes, medicine, or prayers. We want beauty. We also want a sacred place where we can treat our senses to exciting sights and smells and touches. Sometimes we want to play lawn games like croquet, badminton, or cricket in gardens, and at others dine outside, with nature as decor and entertainment. Some people argue that a garden is a work of art, one that's always changing. Others that gardens, however personal, always reflect the temperament of their age. That's true of all creations, of course—one belongs to oneself, but one also belongs to one's time. In *What Gardens Mean*, philosopher Stephanie Ross stresses that

> There is no essential definition of a garden; a garden needn't have any plant material at all. But most gardens do, and accordingly they make statements about our place in and relation to nature. Buildings do enclose us, but they do not, in addition, as do most gardens, make us think about wilderness, other species, interdependence, the passage of time, the limits of control.

She goes on to say that "the dichotomy inside/outside which seems applicable to most traditional architectural structures (for example, houses) doesn't apply comfortably to gardens," and that whereas gardens have boundaries, they're not intended to have a facade or offer "a particular appearance to someone outside looking in." On this I disagree. A garden offers a face, an expression. From the outside it's a mural one can enter. Inside, its details become visible. One can wear it like a voluminous garment, disappear into its sleeves. Insider and outsider experience very different gardens.

All gardens are mind gardens. When poet Wallace Stevens invited visitors to sit in his garden, he liked to tell them how much every bulb cost. I can't imagine how he imagined his garden, or the idea of a garden, based on the cost of bulbs. Alexander Pope, who often wrote poetry with classical themes, extended that penchant to his garden, which was full of classical allusions and illusions. I'd love to know what Dylan Thomas's garden was like. I hope it was as tumbly and lush as his poems.

A garden must be cultivated, worked at. "I am strongly of the opinion," writes the grande dame of landscape architecture, Gertrude Jekyll, in *Colour in the Flower Garden* (1908), "that a quantity of plants, however good the plants may be themselves and however ample their number, does not make a garden; it only makes a collection." A garden is an organism that roams and grows. Like someone you know well, a garden changes over time, changes while remaining the same. Tint the idea of a garden with metaphor and all sorts of innuendos appear. Example: The garden of one's regard. In that metaphor the garden is a pleasure that occupies a special locale in one's life, as does the regard. The garden of a classroom, where young minds are nurtured. A garden of misremembered days, as in "her childhood seemed far away and blurred, a garden of misremembered days." Here the association is with overgrown gardens, but once again the garden (and childhood) is emphasized as a unique locale. In *The Fourth Book of Airs* (1617), seventeenth-century poet Thomas Campion says of (and probably to) his love "There is a garden in her face, / Where roses and white lilies grow; / A heavenly paradise is that place / Wherein all pleasant fruits do flow." In ancient times, in desert worlds, a garden was a favorite spot for romance because few things were more soul-drenching than the idea of an oasis. A hidden garden in the aridity of life soon became a metaphor for love. In the Bible's torrid Song of Solomon, King Solomon sings to his intended that her virginity is like a luscious garden he will soon enter. Then he mentions one by one all the fruits he will pick, all the scents he will inhale.

Thinking about gardens leads naturally to an alchemy of mind. Consider compost. To create it one transforms the waste products of the garden by digesting them (via bacteria) and turning them into rich dark stuff essential for further growth. If only we could cast off life's bad experiences in that way, gather them in a pile at the edge of our awareness, where they're slowly digested and turned into personal growth. "Plants don't point a finger," Anne Raver reminds us in *Deep in the Green*. "If they live, they don't carry grudges. If they die, unless you've killed an entire species or a rain forest, you feel only momentary guilt, which is quickly replaced by a philosophical, smug feeling: Failure is enriching your compost pile."

These are some responses to one's personal garden, feelings not shared perhaps by the keepers of formal public gardens, which span lifetimes like a transmigration of souls. Then there is gardening as thrill-seeking, a special genre as fascinating as it is bizarre.

16

Sometimes people go to extremes and create eccentric, fantastic, obsessive gardens that linger in one's memory and become tourist meccas. For example, the seventeenth-century Spanish Monastery of San Lorenzo de Trassanto maintains topiary mazes so intricate and tightly grown that gardeners can't move among them, but must balance on scaffolding, clipping them from above. There is the Garden Shrine of Fertility in central Bangkok, Thailand, where garden statues are human-sized penises carved from stone or wood. Some of the penises are bowing, some lying down, all are erect, and some have testicles and legs. Women visit the garden to pray for male children, among other less lofty blessings ("winning the national lottery, getting a better job," etc.). Who could resist the Garden of Divorce in San Francisco? Landscape architect Topher Delaney designed the garden as a personal narrative for a client who had run the rapids of a hideous divorce. Among the jagged shards of concrete (taken and broken "from a terrace laid by the client's ex-husband"), some of which look suspiciously like untended gravestones, are plantings of bloodgrass and unsatisfying shrubs. Several prankish surrealist gardens exist, including that of Edward James, a wealthy Anglo-American eccentric who retired to Mexico and put everything into his elaborate fantasy topiaries, sculptures, fountains, and vegetation celebrating surrealist heroes. The encroachment of teeming trees and vines just adds to the effect because in time the entire garden will be reclaimed by the unruly id of the jungle. There's the formal roof garden atop Rockefeller Center, complete with oh-so-proper lawn and manicured box hedges, apparently waiting for royalty and their guests. The cemetery in Tulcán, Ecuador, offers one of the gardening wonders of the world, where topiary craftsmanship is raised to angelic heights. In *Gardens of Obsession,* Taylor and Cooper describe a few of its surprises:

> Several of the hedges have been clipped to create bas-reliefs or architectural mouldings. Others are deeply incised with portraits of South American and Inca heroes or figures drawn from Aztec, Oriental and Egyptian mythology. Finally, single trees have been clipped into colossal shapes, including one of an elephant and another of an overweight astronaut.

One expects to find topiary at, say, Disneyland. However, droves of people with modest homes have also felt the urge to plant topiaries, usually on their front lawns. Most are of animals—giraffes and lions being favorites, with the occasional swan or rhino—but many are more personal, including family symbols like boats, voluptuous mermaids, or even dioramas of events with bas-relief dates. "You will do foolish things," Colette advised, "but do them with enthusiasm."

My favorite topiary is in Old Deaf School Park in downtown Columbus, Ohio. There Seurat's impressionist painting, *Sunday Afternoon on the Island of La Grande Jatte,* has been reproduced as a larger-than-life topiary garden by James and Elaine Mason. The characters in the pointillist painting have been meticulously crafted in wire and covered in shrubs, which twice a year are groomed leaf-by-leaf like coiffures. In all, there are fifty-four human figures, eight boats, three dogs, one monkey, and a cat: the women with their bustles and parasols, the lounging men, the romping children and dogs, even the people boating on the water. To capture the spatial illusion of the painting, they had to be sized and placed in perspective. So the tallest (and nearest) figure is a twelve-foot woman, and the smallest (and farthest) is a twelve-inch dog. It's astonishing. I visited it one afternoon in April and felt like I had fallen into the live painting, where I could watch the people from different angles or join them on the beach. Visitors are meant to stroll with the figures. Instead of dots, there are leaves and blades of grass.

In the Columbus Art Museum, on that same day, galleries bloomed, as fifty floral artists used flowers to interpret paintings, in an annual event called "ART ALIVE . . . Bouquets to Art." The startling results not only looked beautiful, they smelled good; you could study them from different angles, and they had interesting textures. Some artists used flowers to parody or comment on famous paintings. Some tried to duplicate the paintings in a sort of floral realism. Others used flowers to express the mood of the artwork or the sensations the artwork evoked in them. However beautifully they may manipulate our sense of depth and time, paintings are essentially flat and one-sided. The flower art companion pieces added another plane of sensation, one that bridged the material world with the living, leading through plants to humans, sometimes with great wit.

In the eighteenth century, topiaries became so foolish that Alexander Pope satirized them in *Essay from the Guardan* (1713), pointing out that he had a "Correspondent" who "cuts Family Pieces of men, Women, or Children. Any Ladies that please may have their own Effigies in Myrtle, or their Husbands in Horn beam." On behalf of his plant-sculpting Correspondent, he offered a long catalog of possibilities, including:

> Adam and *Eve* in Yew; *Adam* a little shattr'd by the fall of the Tree of Knowledge in the great Storm; Eve and the Serpent very flourishing. . . .
>
> A Pair of Giants, *stunted*, to be sold cheap. . . .
>
> A Quick-set Hog shot up into a Porcupine, by its being forgot a Week in rainy Weather.
>
> A Lavender Pig with Sage growing in its Belly.
>
> Noah's *Ark* in Holly, standing on the Mount, the Ribs a little damaged for want of Water.
>
> A Pair of *Maidenheads* in Firr, in great forwardness.

I'm amused by the idea of topiary silhouettes. At a restaurant in Jamaica, I once found the menu offering "Steak barbecued to your own likeness" and "Chef's bowel salad," both of which promised more than I desired. My leaf-dollies aren't extreme, but they are a bit bizarre-looking, and I'm sure the surrealists would have liked them, especially if I set among them artistic tributes to technology gone mad. I suppose it depends on what you seek in your garden. If it's a serene escape from the newsy world and the worrisome self, then "Even the topiary works of the Renaissance," as Sir George Sitwell observes in *On the Making of Gardens* (1909), "the green ships and helmets, giants, dragons and centaurs, had something of reason to recommend them, for by their very strangeness they would be likely to compel attention, to stir imagination, to strengthen memory, to banish the consciousness of self and all trivial or obsessing thoughts."

17

A hot, humid summer, muggy and sticky. But it's a boon to all creatures that crawl, ooze, or slither. Thanks to the rich soil, flower buffet, and incessant rains, there's a slugfest in the beds. When slugs start climbing up the daylily stems and perching on the petals, when you can almost hear them debating whether to devour the garden now or drag it away for later, when there's more slime around than in a ghost house, slugs have achieved the status of infestation. Just as a nectarine is a naked peach, a slug is a shell-less snail—one that skates on mucus. Thick, gooey mucus from the front foot gives it traction to climb, and watery mucus greases its way on level ground. I envy slugs their eyes, which sit at the tips of rubbery stalks, can be angled into dark corners, and are retractable. When it comes to mating habits, slugs rival the agility of hermaphroditic flatworms, which stand up and duel with their penises. Slugs mate acrobatically, at the end of slime bungee cords, while waving their eyes around. It's sloppy but effective. That drama takes a while, of course, since they're famously slow about everything. On the ground again, they sometimes stand up on tail to scout the terrain. "Leopard slugs," they're called, because of their spotted hide and merciless ways. Mainly water and protein, slugs serve as good appetizers or entrees. Picturing slug stew doesn't make my mouth water, but in some cultures they're a delicacy. In the culture of my garden, they're a plague.

This season for the first time I have resorted to chemical warfare. Gardeners are not fey souls, kind to all living things. We unleash clouds of death. Myself, I'm devoutly organic. I prefer deception, disguise, and disgust as a strategy, and that has always served me well in the garden. But let me tell you it's no field of clover. I barely keep ahead of the plagues, and I still must kill plant-devouring insects by dousing them with soap and water or a sulfurous fungicide. I've offered the slugs three forms of lure: a bowl of beer (they crawl in and drown); a terra-cotta volcano baited with sweets; or—my personal favorite although it doesn't seem to be working—copper collars wrapped around the base of precious flowers to create a simple wet-cell battery that gives slugs a shock. Citrus growers in California have solved the problem through cannibalism. They employ a tribe of related but far more picky snails (the Mediterranean decollate), which feed mainly on the eggs of their common garden (European brown) cousins. Homeowners can buy the decollates from Bio Pest of Oxnard, California, fifty snails for $10. They arrive dormant, like space travelers in suspended animation, and when you splash them with water, they wake hungry and ready to rampage. Should I stage a slug civil war in my garden? Or, rather, could I do it without imagining the carnage underfoot?

I'm writing this while seated on a cement bench created by stone masons in the nearby lake town of Skaneateles—"Scanty Atlas" as we locals say. A solid perch, its beauty lies in the slab seat, where colored glass has been crafted to make a mosaic of purple roses with green leaves against a background of mottled pastels and earth tones. When you stop to think about it, it's strange that instead of sitting on the ground, humans like to pulverize the earth and shape it into something else, only a yard higher, which we then sit upon. We don't exactly lord over nature by increasing our stature those few feet, but there's a rise in status and illusion of control that's appealing. Especially when nature seems to be setting off fireworks.

Overnight, brilliant sundrops have begun to open, and the yard is blazing with the great waves of yellow flowers I associate with June. This year the sundrops are three feet tall and thick as wheat. Inside the rose garden, forget-me-nots have grown spindly, and it's impossible to leave the rose garden without an encrustation of fragmented seedpods—tiny green balls that stick tenaciously to one's socks. A fun thing to do with kids at the end of the summer is to let them run in their socks through a meadow, and then plant the spore-and-seed-clotted socks in an unlikely spot, where the following spring a wildflower meadow will sprout. Children love to play in a garden, but they like to have their own kid-sized garden, too. A sort of petting zoo, it welcomes them to the wonders of nature and the simple intransitive pleasures of merely tending. In *Candide,* Voltaire advises readers to seek peace tending their own gardens. There is an ecology to every life, and each family is a garden, where sensitive family members grow in varying degrees of harmony.

Every flower is a garden, a small Eden for a nectar-craving butterfly, bee, or bird, not at all the petaled lushery we rub against our senses for enjoyment. Its goal is far more brazen: voluptuousness is its job, deceit its genius. A virtuoso of bribery and disguise, it thrives on conning unsuspecting travelers. Flowers get all gooey and colorful to tantalize the likes of a moth, ant, bat, bird, or mouse, in the hope that an agile stranger will move their pollen around. Since they're mainly stationary (except for the occasional bend or wobble), they need to persuade some other life-form to perform sex for them. Their story is as honest and unsluttish as that, reminding us that nature doesn't judge or mean; it just is. *We're* the creatures who strive to be modest, moral, or merciful. The rest of nature mainly struts, exposes itself, and oozes. Mind you, sex is probably more fun for us than for a flower—contrary to what some well-meaning folk may protest, I'm not convinced flowers experience a mood like fun—but it's also more torturous. An open secret among humans is that at least once in one's life, one comes to wish sex were as straightforward for us as it is for flowers.

Although flowers convey volumes of information to bugs and birds and one another, humans have always employed them as vivid messengers of love's many moods, and the Victorians went so far as to treat them like servants, making sure they dressed properly, spoke correctly, and had all the right references. Sending your sweetheart peach blossoms meant "I am your captive." Sending him aloe meant grief. A bright jonquil meant "Return my affection!" Bluebells were for constancy, but buttercups shouted ingratitude.

Because pine needles spelled pity, one urgently noted in which direction they were tied. If the flowers arrived with a ribbon tied to the left, they spoke of the sender; tied to the right, of the recipient. A petunia meant "Never despair," while yarrow threatened all-out war. One might compose a bouquet of peonies ("Feeling bashful"), zinnias ("Thinking of you"), burgundy roses ("Unconscious beauty"), and snowdrops ("Hope"), in what amounted to a visual haiku. Lovers have always looked for ways to sidestep the embarrassment of revealing their gaping, ravenous, unadorned need. Sending a flower code was less risky than declaring one's feelings by blush, behind trembling eyelashes. The feeling need not be a surprise to the beloved, of course, merely heretofore unstated. How like a flower unfolding in exquisite slowness to a rapture of scent, form, and color is a deepening romance. When fully open, a flower offers no other possibility. It is the full expression of its destiny, the culmination of routine biology and individual traits. When fully revealed, love seems always to be a fateful crossroads where two quirky people arrive hauling all sorts of baggage.

Flowers are like poems. Consult them for delight, and they'll delight you. Look to them for deeper truths, and you'll find much to mull over. Some of their music falls on deaf ears, some of their billboards go unseen. For example, I would need the eyes of a bee to spot the target-shaped designs on many flowers. Providing perfect approach guides and landing pads, they invite a militia of rowdy visitors to come and dine, in the process smearing them with pollen to carry blindly to the next blossom. They do not know their offspring. But, then, they do not know. They simply are. Given that, how lucky we have senses that relish colors and savor fragrances. We don't need to be attracted to flowers. That they delight us is a happy accident, only a lavish form of sensory noise, a casual byproduct of evolution. All the more reason to peer into their nooks and crannies, convolutions and folds, hanging-out tongues and plump pouches. If they "hide their light under a bushel," as the adage goes, then they are both light and bushel, the beginning and the end, the blooming dance of life. They are megaphones of scent, broadcasting their whereabouts in molecular slogans. We color them with our memories. They stain us with their charm. We gather them like bons mots on a sleepy morning, and arrange them face-to-face in vases, as if they could speak among themselves. A guest in the summer house of the soul is a flower. Gardeners love to page through catalogs in which flowers pose nakedly in darkness and light. They serve the viewer's eye but elude the touch, which is sometimes the way with beautiful things. But the silky pages of a catalog may be stroked like petals. Floral shapes, glimpsed in half-light, can illuminate paths through a garden of memories. And when one applies the ear trumpet of imagination, flowers can be heard whispering some of life's oldest secrets.

18

Persis and I swim at daybreak, following our usual routine of twenty pool lengths, then pausing to chat before swimming twenty more, pausing again, and swimming a final ten. At the first pause, we notice a curious phenomenon: swords of light twirling above the water. It is exactly 8 a.m. and the angle of the sun is such that light reflecting off the agitated water onto the steam creates dancing beams. Why straight beams? Because despite how soft and rounded the water looks, it's made up of many planes. One often sees streaks of light projected on the sides of pools, or lozenge-like shapes in the surface of the water in bright sunlight. But this is the first time either of us has seen beams dancing out of the steam and swirling in the air. Swimming pool physics. When the apparition stops, Persis swims halfway up the pool to agitate the water, and sure enough the beams return.

Another twenty lengths, followed by a pause, during which we chat, as we do, about everything and nothing. Today, I'm eager to get her opinion on something fascinating I've read. Personally, I don't talk much to my plants. Chrys, on the other hand, feels that talking inspires them. But how about playing music to them? Music is organized vibrations, and living things respond to vibrations, so could favorable vibrations stimulate plants (or people) to grow? Joël Sternheimer, a French physicist and composer, argues that it's possible to reproduce in music's audible vibrations the inaudible quantum vibrations that happen when amino acids combine to make a protein. Notes correspond to the amino acids in a protein necessary for plant growth, he explains, and it's possible to combine notes to make the sound wave equivalent of a complete protein. "Each musical note is a multiple of original frequencies that occur when amino acids join the protein chain." When a plant hears the right molecular tune, it doesn't perceive what we call music but rather an instruction to synthesize more of the protein, and in doing so it grows. According to Sternheimer, tomatoes grew over twice as large as controls when serenaded by six growth molecules three times a day. He's just as interested in musical inhibitors. Playing inhibitor music, he claims, can stop the growth of an enzyme essential to a virus that's making a plant sick. His scored melodies are patented for use in agriculture, but also for textiles and health. But "don't ask a musician to play them," he warns. "You must be very careful." If they can affect plants, they can affect people, too, in ways not yet fully explored. Plants take in carbon dioxide and give off oxygen; humans do the opposite. That makes us well-matched, but our needs and responses are very different.

Also a physicist and musician, Persis seems like an ideal person to ask about Sternheimer's theory, and there's no doubt she's intrigued. All the world interests her, especially the delights of watching and helping her three very different children grow, doing the nuclear physics she's known for, playing chamber music on her cello.

"The problem is, the wavelength of sound that you can hear is much larger," she says thoughtfully. "Sound is like an oscillating wave, sound is like a ripple in the air, which has crests and troughs and crests and troughs. And the distance between crests and troughs is much bigger than any size of any molecule. The scale seems wrong. If you were to put in really short wavelengths, you could do things to molecules, but these are just too long. The inaudible quantum vibrations are at a totally different frequency from audible vibrations—orders and orders of magnitude different—so it's hard for me to understand how one could affect the other. It's like saying that something that affects the solar system by moving a planet a little bit is going to do something to our neighborhood, or that something that happens in our neighborhood is going to affect the planets. Sounds screwy to me."

I knew she'd find it amusing. In the seventies people were saying that if you talked to your plants they'd grow faster. These days appealing to scientific authority is more popular, especially since most people know so little about the intricacies of biophysics. That's why I often ask Persis about the latest scare or promise. This protein-concert theory is novel, and I'll leave it to researchers to test. Off we go on another set of laps.

At the next pause, unexpected voices flow through the gate—her engineer husband Jim with their littlest, Rose. Last night, when Rose and Persis were out walking, they came upon a rosebush that delighted the child. Could *they* grow a rosebush? asked Rose. Mom explained that roses are hard to grow. "But you're not hard to grow," she had added, and five-year-old Rose thought that was immensely funny.

Then Persis said that I had a rose garden they could visit. Rose was excited. It must have stayed on her mind, and this morning, when she knew her mom wanted to swim with me among the roses, she thought she'd join in.

Our swim finished, I give Rose a tour of the garden beds, sprigs of lavender and peppermint to play with, and three roses to take home, which I clip on very short stems to avoid the thorns. I choose ones that smell good: clove-scented Dainty Bess, lemon-nasturtium-scented Peace, and musky orange Singin' in the Rain. I invite Rose to visit me and the roses any time, and we all walk across to her house, pausing briefly to admire a beautiful moss glade.

I have flowers that have been alive longer than five-year-old Rose has, trees that greeted men returning from the Civil War and, before that, Indians perhaps. Some people find such thoughts disturbing. How dare plants outlive us? Myself, I love the way trees provide a visual bridge between earth and sky. I feel cozy among trees, not suffocated by their overlapping leaves, or deprived of sunlight.

I guess it's hard for people to understand the quiddity of a tree, whose purpose does not include us. Yet we share similar instincts when it comes to survival. They may not experience what we refer to as pain, but they recoil from damage. They may not feel parental, but they'll kill to ensure the continuity of their genes. When I say "kill," I mean it literally, brutally. Plants are not mild-mannered. They retaliate with impressive chemical warfare. They devise hideous poisons like strychnine and atropine; ghoulish irritants like poison ivy and poison sumac to blister the skin; spines sharp as hypodermics. Even spores, seen under a microscope, look deadly as medieval weapons. They're spiky balls designed to cling and annoy, so that they'll be carried far and released. Small wonder they irritate our sinuses and eyes. Holly and thistle have sharp spines, blackberries and roses have curved thorns (which also help the plants climb), stinging nettle hairs work like syringes loaded with such irritants as formic acid and histamine.

Three passionflower vines, thick with blooms as showy as military regalia, are climbing up trellises on my back patio. They're tropical plants, happy here in summer if they're watered each day, but I would need to use a systemic insecticide if I took them indoors for the winter, so I'll probably give them away to Chrys and Bill at summer's end. Even their leaves look beautiful, but if you decide to add some to a salad you'd best beware. Passionflower leaves contain cyanide, which is released only when something bites into them. Then a cell wall breaks between two otherwise harmless substances, which in combination become lethal. This works the way nitroglycerin does, or, on a less alarming note, the way table salt does, or the way fireflies flash (by mixing luciferin and luciferase). As might be expected, because nature is often an arms race, some leaf-eating caterpillars have evolved an immunity to cyanide.

Like animals, plants sometimes find strength in numbers: if danger threatens, they can warn relatives. When an elm is being attacked by insects, for example, it releases a chemical that alerts others in the grove to produce poisons.

We happen to find some plants like rosemary tasty, but the flavor didn't evolve for our pleasure; it evolved to torture the noses of animals, which would then keep their distance. The same is true of sage, thyme, lavender, peppermint, and many other herbs. Perverse as we are, we enjoy their piquancy. But to most insects and animals, they deliver pain or disgust. We were never part of the plants' useful universe. We just appeared one day as a sort of aggravation. Plants can employ us the same way they employ deer or lions—as chemical messengers easy to bribe and deceive—but they don't really need us. In fact, they did just fine for millions of years without us. So I feel we're privileged to walk among them and find pleasure in their ways. It might have been otherwise.

19

Paradise has returned. Not minimal paradise, mind you, not the bare bones of a paradise too weak to count on, but a tough resilient paradise with thick strong limbs, erect posture, and vigorous ways. A thorny paradise. There's nothing delicate or fey about its blooming fists of lavender. "Yikes!" I say out loud, in a mixture of delight and confusion. It's like waking up to find your dead uncle returned as a heavyweight boxer standing in your garden.

A matte lavender rose sometimes edged in dark purple, Paradise is not a keeper in my locale. All of the purple and lavender roses fare poorly here, where something in their lineage makes them vulnerable to our winters. But with the weather topsy-turvy now, you never know what curio might appear to defy custom and turn local wisdom on its ear. I have scant success with the lavender Angel Face, but perhaps I'll give it another try. Unlike Paradise, Angel Face has golden majorette fringes inside when it opens, and petals delicate as veils of rain.

If I were a little girl, I suppose, I might arrange roses in a semicircle of bud vases and pretend they were dolls, because they do have faces, just as books have faces. One concern I have about the current generation of electronic books is their facelessness. Books look different, and that adds to the pleasurable illusion of carrying with you a distinctive mind. That may be remedied one day in e-books, but for the time being I respond to the personality of bindings and dust jackets. I've written at length elsewhere about the natural history of the face, and why we're driven to see faces everywhere, even on rock formations on Mars! Suffice it to say that we like to inhabit the nonhuman world with recognizable friends and fiends. I bring this up because people are sometimes chided for what's called "anthropomorphism," attributing human characteristics to nonhuman plants and animals. We've inherited that prohibition from the outdated belief that we are not animals and therefore share nothing with other animals. That leads to the assumption that we can learn nothing about human behavior by observing plants and animals. How arrogant and how foolish. Evolution has used many of the same tricks for plants, animals, and humans alike.

I love sitting at the crossroads where nature and human nature meet and each throws light upon the other. So although I don't imagine my plants share human concerns and emotions, I do respond to their unique faces, as well as to their motives, strategies, culture, and health. Do plants have motives and instincts? Absolutely. To the best of my knowledge, they don't have consciousness. But they *are* self-aware. They know when they've been hurt, and they can take stock of their circumstances and adjust their behavior. For example, on frosty spring mornings, I often find tulips bent over, as if dead; but they're up again by noon. Freezing water would explode the cell walls, so to protect themselves, they go limp when the temperature plunges and send water down to their feet. Then they shoot water back up to the flowers when it's safe.

In times of drought some animals become nocturnal to reduce their need for water, while others—say, fish in a dried-up pond—go into a sort of hibernation until the rains come. Most plants go limp. Flowers tell you when they are thirsty. Few things are as pathetic as an impatiens shriveled up and drooping like a spaniel with its muzzle on its paws. Dry soil can leach the water out of plants, which then roll up their leaves to conserve moisture. They have a better chance of survival if they sacrifice leaves and flowers and pull all the sustenance down into the bulb or roots. They can grow new flowers and leaves, but they can't afford to lose their core. Odd though this might sound, evolution equipped women with a similar response, which I learned about in the Antarctic, whose waters are so frigid that you will die almost instantly if you fall into them. Women are in more danger than men from hypothermia because women's bodies preferentially protect the reproductive organs, and will pull blood from the brain, heart, and everywhere else to warm the reproductive core.

Plants grow feverish when they catch a virus, and they even get the equivalent of flushed cheeks. Researchers from the University of Ghent infected tobacco plants with tobacco mosaic virus, and using a high-resolution infrared camera, discovered that leaf temperatures were higher at the sites of infection, and that the fever appeared before any visible signs of illness. The discovery will help in the early diagnosis of crop disease, but it also adds to our understanding of what we have in common with plants. They get feverish when sick, droopy when under stress.

An exuberance of roses this year. So many are blooming that I gather three dozen to bring indoors, and only stop because I've no more room in my pail. I don't remember a year when roses bloomed this lavishly. The Colette rose (firm buds like young bosoms, opening to light pink blooms with a wash of apricot) is towering up the fence, creating a flower trail as it goes. The real Colette, who wrote books in which she treats flowers and people with equal respect, would adore the fountain of pink flowers—and the unusually large thorns, too. Along one fence, four climbing rosebushes rearrange their blankets of color each day, as some blooms wither and others open. Reine de Violette, an old-fashioned bushy purple rose that unfolds into a shallow ruffle the color of grape juice, has become an avid climber for some reason, and begun streaming over the fence. Next to it a pool of deep pink roses, then a spread of orange-red roses, then an explosion of small, pale pink roses, and finally a huge flight of blue-red roses growing in bunches of five to ten buds. Behind them, a rose the darkest red I've ever seen climbs a tall black obelisk. What a display. Noses (the people who invent perfumes) refer to floral scents as notes, their combinations as chords, and in their workrooms they tend to arrange scents like the keys of a pedal organ. In the scent-music of my rose garden, there are major and minor chords, simple notes one can smell for a long while, and others the nose smells fleetingly, which seem to evaporate like sixteenth notes. Every rose has unique qualities, opens at its own pace, and grows into its color and scent in predictable ways. Weather is a wild card, as are flukes of breeding, so just when you start to take a rose for granted—ho-hum, another rose equivalent of a trapeze artist—nature grabs you by the lapels and wakes up your dozy senses.

20

A strange sight in the garden has been alarming me all week, and today it's unbearable: twenty roses with broken necks. Climbing Eden and Heritage have produced many huge cream-and-pink blooms, each one densely petaled and heavy. Alas, their stems are weak, so just as the buds open fully, they break their necks! A breeding error. Let your head get too big and see what happens. What to do? I make a mental note to move them next year to a fence where I can tie the canes securely, but even then I may need to brace each flower.

So on my garden rounds this morning, I carry metal hoops, wooden sticks, and green twine. After trussing up what's left of the Climbing Eden and Heritage blooms, I strut gangly stems of Jerusalem artichoke, whose sunflower heads have begun to fall and break. The sprawling pink-and-chalk blooms of Carefree Beauty have toppled onto the phlox, smothering it, and I hoop in some of the rose's wayward branches. A mound of giant daisies, a scant week or two from flower, is tall and straight at the moment, but I know from experience they will flop onto the grass and block the garden gate. Another hoop skirt. That done, I pick up several tools, head for the weedy tangle behind the apple trees, and quietly begin tending the glade of primroses and wildflowers.

We think of apples as wholesomely all-American, but our apple trees are immigrants that arrived in the seventeenth century and spread rapidly, thanks in part to the generosity of Jonathan Chapman (aka Johnny Appleseed, 1774–1845), who handed out apple seeds to everyone he met. He owned 1,200 acres of apple trees himself. The apple is an ancient tree in the rose family, native to the mountains of western Asia, cultivated since prehistoric days. Egyptian pharaohs grew apples along the Nile in the thirteenth century B.C., and the Romans finally introduced them to Europe in the third century. Good to eat, they're also good to smell, and I enjoy thinking of ancient Egyptians napping in the fragrant shade of apple trees. A few gangly raspberry bushes remind me of my grandfather, who loved to garden and to sleep outside while others snoozed indoors. Hot work, tending a primrose path. When I'm finished, I hang up my cotton gloves, and survey the lawn for a soft piece of earth to lie down on.

Few things are as wondrous as sleeping outside on a star-loaded summer night, as planets glow steady as lighthouses, shooting stars sigh toward the horizon, and a carnival of sky exotica like black holes and white dwarfs lurk among the constellations. If you're far from civilization, in a well-like darkness rarely known to city dwellers, more stars will be on view than mind can comprehend or astronomer tally, a grainy encrustation of lights. Thanks to the recent Hubble telescope photographs of lusciously colored nebulas and sparkling star nurseries, it's easy to imagine the riches just out of eyeshot. Without visual clues, the mind plays tricks, and it can feel like you're tumbling upward.

Most people lie down at night and open the dream gates for a while in sleep's bizarre time-out, fret factory, and repair shop. We don't happen to see well without light, so the rest of the world seems to doze when we do. But it's peak time for night-dwelling animals like bats, foxes, coyotes, and bobcats. The next shift of Earth's residents—the night-working day sleepers—is busily on the prowl, and sometimes you see the tiny red eye-shine of spiders when you aim a flashlight at the ground or hear owls hooting, or bats fluttering, or raccoons climbing up a tree. Sleeping under the stars always gives me a sense of perspective—the small scope of human doings in the richness of nature and the vast sprawl of the universe.

I've slept outside in different seasons, in different time zones, and at different altitudes, but one of my favorite times was on a schooner off the Bahamas, spending a couple of weeks studying spotted dolphins. Belowdecks, the air was fetid and oily, so I took my air mattress topside, sprawled out on the deck, and was soon cradled to sleep by the gently rocking boat. In the night I woke to find that I had somehow slid across the deck and was stretched feet first halfway over the side of the boat. Laughing out loud, I simultaneously got a chill when I thought of waking up disoriented and edible in the water. That was enough to send me belowdecks again, but not before sitting a while to enjoy the stars and moon reflecting on the water in great pools of frying light. The moon looked bright as a neon sign.

Between snoozes in my backyard, I watch the moon rise and think of the astronauts who strode across the Sea of Tranquillity thirty years ago. *We've been there,* I say to myself, still surprised all these years later. When Neil Armstrong shambled onto the moon's stark plains, it became the first of many stepping-stones into outer space. Indeed, he left a human footprint in moondust in the Sea of Tranquillity. What a jolt that landing gave us. From telescopic scrutiny like that done by John Whipple in 1851, we knew the moon was pockmarked with dry craters once thought to be thriving oceans, but little more—not even what landing on the surface might entail. Would a human sink into it like quicksand? Would a spaceship? Would there be signs of life? Our deep cosmic loneliness drove us, as it does now, to keep searching for others like us, creatures with a fever in the cells, moving beings in a universe of blanks, a life-form that quests.

It was not a short trip: it took humans a few million years to get to the moon. Or around 15 billion if we begin at the beginning of the universe, a modest ball of hydrogen all in one place and solid, which exploded in what we whimsically call the Big Bang. Our atoms were forged in that early chaos, and the rest is history. Fast-forward from those atoms to a human being pressing his foot into lunar soil.

Of course, it was disappointing to discover the moon's searing bleakness. Imagine something that big being dead. But it was also thrilling to watch our kind landing on another world in a confetti-shower of light, a one-man parade. I remember devouring the scene over and over on television, in joyous disbelief. The real breath-stealer when the astronauts turned the camera back toward Earth, floating like a milky-blue marble against the jeweler's velvet of space. The thinnest rind around the planet contained the ground, sky, weather, and all of human history, everyone we knew or loved—our whole experience of life in one place on a small, fragile sphere. That picture from the Apollo 11 mission told the whole story and helped us understand for the first time how seamless the planet is, what we share with our neighbors, and a little of our real address.

According to legend, St. Francis was strolling through the lanes of Assisi one evening, enjoying the cool compress of the night air, when the moon rose like silent thunder, a giant golden coin. Its light cast pools of glitter on dark streets and cut unseen worlds from invisible landscapes. Beholding this marvel all alone, he ran to the church and rang the bell with wild delight. People rushed outside in alarm, because church bells rang at night only in times of tragedy, warning, or disaster. When they arrived at the church and saw St. Francis in the bell tower, the townsfolk anxiously called up to him to tell them what was wrong. With urgent enthusiasm, he answered: "Lift up your eyes, my friends. Look at the moon!"

Some nights I too have wanted to wake the neighborhood to see the moon at the end of the street, especially when it's perched like an owl on a tree limb, harvest-huge. Figuring heavily in myth and lore, the moon has tantalized human beings because it peers down at us like an ever-present companion, controls the seasons of animals, as well as the give-and-take of the oceans and our inner tides (the word "menstruation" traces back to the word for moon). The night-light of Earth, the moon has shone a path for land trekking and guided seafarers home. It has foretold the weather and helped us frame our elaborate ideas about time. We still have much to learn about its pull on us. As Wallace Stevens slyly notes, "The book of moonlight is not written yet."

Some gardeners consult the moon for guidance about the ideal time to plant or harvest, fertilize, take cuttings, and even weed.

Just as it controls the to-and-fro of the tides, they argue, the moon's waxing and waning influences plants by pulling energy above- or belowground. The general principle goes like this:

> As the moon wanes, from the full moon to the last quarter, plant anything that requires a large root network or underground storage system—like bulbs, tubers, biennials, and many perennials. . . . During the last phase of the moon, from the last quarter to the new moon, don't plant anything. Instead, pull weeds. . . . The best planting conditions occur when the moon is in an appropriate phase and in the house of the zodiac sign that is considered fruitful, such as Cancer, Scorpio, Pisces, Taurus, Capricorn, or Virgo.

Lunar gardening may sound *au courant* and New Age, but it has an ancient history, especially among astrologically attuned cultures. Even *The Old Farmer's Almanac* abides by the phases of the moon to create its gardening calender. Personally, I follow this simple rule of green thumb: Unless you're living off the land or the land is your living, the best time to do gardening chores is whenever you're in the mood. You'll win some, you'll lose some, and some will get rained out, but at least you'll enjoy your garden.

For a moment a galaxy seems to have fallen to earth, but no, it is only a cloud of fireflies dancing on the lawn. All year I wait for their brilliant intrigues and one of the great treats of summer nights: watching auroras of fireflies. I've sometimes caged one in my hand and inspected the small lantern of its body, flashing a cold green light, one of the brightest cold lights ever devised. You can read by it. Our hot lights waste energy, but firefly light is almost perfect.

Both males and females flash, using a personal code during courtship. There are also *femme fatale* lightning bugs, which lure other females' mates by mimicking their passwords. I say "passwords" as a kind of shorthand, because the real code is an absence, not a presence. A male flashes and waits for his female's single-flash reply. How long she delays is what the male deciphers. Other females know this. Then, when a male dives down, expecting to mate, the *femme fatale* eats him, acquiring a chemical he carries that will arm her and her offspring against predatory birds and spiders. What we see is a night festooned with their private dramas of love, loss, danger, and display.

Eighteenth-century American women used to tie fireflies in their hair as decorations—sort of living tiaras—but I'm not sure how they fastened them (with thread? with individual hairs?). What a sight that must have been—a garden party by fairy light. Especially if it was held in a moon garden—that is, a garden planted with white night-blooming flowers. By day such a garden looks talcy and cool, and a little formal with its pale flowers framed by green leaves. But at night it attracts nectar-gathering moths, and the moon reflects off leaves, spiderwebs, and perhaps the water in a well-placed pond. Add a dose of fireflies, a few Japanese lanterns, some friends, and stir gently.

Under the starry overhang of the night, I'm reminded of these lines by W. H. Auden: "Out on the lawn / I lie in bed, / Vega conspicuous overhead." Accentuating the magic, a band of tree frogs, leopard frogs, toads, birds, and crickets offers a strange, syncopated serenade as they fill the night with their pops, plucks, rattles, rasps, buzzes, and moans. They're all speaking different versions of the same word: summer.

21

Sitting at my desk one sultry morning, I'm distracted by beauty. Parked right in front of me is a teal vase containing a garden bouquet of twenty roses, a bright spectrum of colors: fuchsia, blue-red, talcy pink, peach, yellow-white (the Swan), purple-and-white stripe (Purple Tiger), orangey-cream tinged with pink (Abraham Darby), and a hot pink that verges on red (Leonardo da Vinci). They all have names, of course, such as Intrigue, Autumn Sunset, Lasting Peace, and Fame. Some open simply, others are densely ruffled. Some roses seem to be tumbling from the squat round vase, tumbling yet stopped in midair. Others are braced by the lip of the vase, chins resting. Most are fanned out at different angles. But for the green foliage giving them a context and weave, they would seem to be floating clouds or a rose iceberg of different hues. The whole effect is beautiful enough to make one cry out, not in pain but in beauty. Wow! is as close as English lets us come, but it falls short of the sensory stun of bold saturated colors and shapes sparklingly different and unexpected. Ralph Waldo Emerson wrote that there were times when nature made him glad to the brink of fear, and that captures it much better. Although everyone feels rivulets of wonder, and even bone-shaking awe, from time to time, not everyone is as comfortable expressing those feelings as freely as John Muir, Ralph Waldo Emerson, or Walt Whitman. I suppose what people fear is loss of objectivity. But life doesn't require you to choose between reason and awe, or between clear-headed analysis and a rapturous sense of wonder. A balanced life includes both. One of the fascinating paradoxes of being human is that we are inescapably physical beings who yearn for transcendence. One can be spiritual without believing in a supernatural being. Most often, the result is simply heightened emotion of a commonplace, gee-whiz sort. But someone like Lewis Thomas, for instance, had a real gift for conveying his sense of wonder in uniquely artful ways. Unfortunately, language really stumbles when emotions surge. So we don't have a precise vocabulary for complex feelings. Small wonder people resort to metaphors to express their raw joy, mysticism, and awe. There are moments when, as the poet Rilke puts it, "I would like to step out of my heart / and go walking beneath the enormous sky."

Anyway, the roses are blooming like a meteor shower, thicker and healthier than ever this year. Not just for me. My neighbors' gardens are luxuriant, too. Nodding peonies, unexpected blurts of dangling wisteria, thickets of oregano and Russian sage are taking over the neighborhood. But my roses are miracles. Carefree Wonder—pink with a talcy white back—is so bushy and bloom-laden it obscures a large window. The climbing red Blaze finally got the knack and is arching hugely on a trellis with a cascade of blooms. Red passionflowers and purple sweet peas climb the same trellises, but lower down. I've been able to make whole bouquets of one rose: eight Abraham Darbys or a dozen Dark Ladys. Most days I gather two to three dozen roses. Soon they'll rest some, and I'll mope until they flush again later in the summer. But at the moment it's a rose riot.

Choosing among flowers may be like choosing among children; each has its delights. But roses are the main flowers I bring indoors. My passion for roses borders on obsession. I have 120 rosebushes, and I feel that if I were to count them every day, *that* would be an obsession. I tell myself that at the moment it's just a hungry penchant allowed to run wild. Even though I have a splendid array of rosebushes, I keep buying more. I can't resist a healthy, heavily budded rose, and I rarely let the blooms stay on the bushes. The second they're ready to pick, I bring them indoors for arrangements. There are four dozen cut roses in the house as I write this, and more are coming into bloom each day, almost more than I can handle. I am besotted with roses—and I bought two more rosebushes in the middle of a bike ride this morning. My ride takes me past my addiction, two nurseries where my delicately petaled and perfumed drug is sold.

I'm not an expert rose person by a long shot, but when I go to garden shops, people often assume I'm an employee and ask for advice. I think it must be because they see me checking the roses so carefully. Does this one have mildew? Does that one have black spot? How many buds are there? How much new growth? To my surprise, this year I seem to have acquired enough experience to be reasonably helpful. Most often they ask for fuss-free, surefire roses anyone can grow, and I suggest the pink Fairy, a hardy rose that blooms repeatedly until frost in my region, and which, for some reason, the deer don't eat. It has small flowers among the thorns and perhaps deer feel the blooms aren't worth the skirmish. I also suggest the Carefree roses—Carefree Beauty, Carefree Delight, Carefree Wonder—which handle diseases well and grow quickly into large luxuriant mounds of pink. Another good choice is the English tea rose Abraham Darby (named after a nineteenth-century British industrialist), whose scent is gorgeous, whose ambiguous color (a lightly stirred mix of apricot, pink, and yellow) is spectacular, which opens densely petaled, and blooms generously all season. If they like purple, Reine de Violette is just about the only purple rose I've found that's reliably hardy here. It has an unusual growing habit—very bushy with lots of leaves, among which sit many flattish, fluffy, grape juice–colored flowers. I think it looks more like a fruit bush than a rosebush. Blaze is the most popular red climbing rose for trellises, and, although it takes a few years to rise high and get solidly established, it produces fountains of red flowers. Old Blaze roses decorate many of the porches downtown, as if the homeowners have draped red flags from the roof. Othello is another good choice for a tea rose. It's sturdy, hardy, and can climb (one of mine is twelve feet tall). The flowers are huge, magenta, loaded with heavy scent. At the moment, my tall Othello, climbing the fence by the bay window, has fourteen large blossoms. People are often tempted by Don Juan, a dark red tea rose that's smolderingly beautiful and seductive. But I've found Don Juans to be delicate and easily shocked when transplanted. And you absolutely cannot pick one of the flowers until it's ready. Even then, it might not open fully if you pick it. Another red rose, the Squire, is a little more responsive, but also can be stubborn. Pick these roses when the petals begin loosening, and they won't open indoors. I'm not sure why. They get water to drink in a vase; their stem is a straw for water on the plant. But they also need a chemical that signals them to flare. To be safe, with Don Juan and the Squire, I wait until the outer petals have pulled free and are admitting the air. If a traditional long-stemmed red rose is what they want, I suggest they try

Chrysler Imperial or Mister Lincoln. Another favorite, which looks like it should be a real troublemaker but isn't, is the red-and-white-striped Scentimental, a rose with an unusually heavy scent and a festive look. But like most roses, these last five require some care. I could fill an entire chapter with favorite roses I find resilient and beautiful. I've bought several Canadian climbing roses, and I'm pleased by how well they survive harsh winters.

The more we know a thing, the more we appreciate its subtleties. I love my roses for their shades of difference, this one's sepals, that one's pink (say, Leonardo da Vinci's, only a hair's breadth more puce than Jean Piaget's). Such small shadings create individuals, perhaps with very different habits. Roses are *carpe diem* flowers—I'm never sure which ones will survive the winter. It's like adopting many children without knowing if they'll thrive. I appreciate them all the more for not knowing if this will be the only season I get to cherish their beautiful faces and fascinating ways.

Rose, where did you get your red? While taking my vitamins each morning, I sometimes pause at the lustrous oval gel of mixed carotenoids. If I took enough of them, I would turn orange; one can buy carotenoid pills that produce a suntan. Carotenoids give rose petals their red and fall leaves their shimmery scarlet oranges. I recently learned that roses have an enzyme that transforms carotenoids into the scent we associate with roses. This is just another example of nature's using one chemical in several ways. "How many things can you do with a pencil?" one might ask people, to encourage creative thinking. Evolution has been playing that game since the beginning of life. I would have said since the beginning of *time,* but of course life existed before time, a useful idea which is our own human creation, which we obsessively calibrate in terms our senses can fathom. Carl Linnaeus, the eighteenth-century botanist, even invented a floral clock. Noting that petals open and close at the same time each day (because they're synchronized to how much light they receive), he arranged flowers in sequence, using the movement of petals to tell time.

Variety is the pledge that life takes, and it doesn't apply only to chemicals but to behaviors as well, especially ones related to mating and child rearing. "Waste not, want not," an old adage goes. Redundant and wasteful as the brain can be with its cells, it also has evolved ways to use the same pathways for different tasks. Different trains travel on the same tracks. Not only brains use this strategy, flowers do too. Chemists have been diligently searching for the rose's aroma enzyme, because it's hard for perfumers to produce pure floral notes in a laboratory. Or for some flowers to produce them at all. When growers create the dazzling inbred garden plants we love, they often sacrifice aroma. Flowers forced to use their energy on color have little left for making scent, which means they don't attract as many pollinators (except humans). Thus my beautiful, tall, paint-splatter snapdragons are scentless. But a molecular biologist at Purdue just isolated the scent gene of snapdragons, and she plans on reinstalling the gene in de-scented flowers. Don't imagine this is an idealist's attempt to restore what was lost; she's also thinking about accentuating the scent of some plants. "If we can increase scent, for example, to apple trees," she explains, "the flowers will attract more pollinators or recruit new foragers. And that means better fruit quality."

I'm told fragrances sell more in summer than any other time of year, and that doesn't surprise me. All of nature is in bloom, oozing scent, lush with romance, and our instincts tell us to join the party.

When rose petals fall, the house looks littered with bright paint chips. They fall in slow motion and soft as rain. I like to lie on the couch with the back door open and two or three vases filled with roses on the coffee table beside me. The roses create a smell veil between me and the outside world. They bathe me in scent each time a light breeze touches them. In between breezes, they're still and my mind shifts to some other awareness. Then quietly, subtly, the roses begin mumbling scent once more.

22

On Saturday mornings, the farmer's market on the lakefront offers a bazaar of flowers, foods, entertainment, and curiosities. Leaving it to formal European spa towns to host promenades of the lavishly dressed, locals stroll in Birkenstocks and jeans to meet old friends and gab with acquaintances. It's closest, I suppose, to an impromptu cocktail party with health food and crafts. I often go there to buy annuals, hanging baskets, or a rare perennial, but mainly I love to look at the museum-quality eggplants and melons, the sweet corn so fresh its silk threads are still glistening, the just-picked strawberries you can smell from yards away. I relish the sensuous beauty of perfect fresh vegetables laid out as a mosaic of rich color, fragrance, and texture.

Local growers sell their wares, and craftspeople also have booths. Only handmade or hand-grown items may be sold from the market's wooden stalls. At either end of three long promenades, or at the two intersections where they meet, musicians play everything from bluegrass to glass harmonica. Sometimes morris dancers, belly dancers, or one lone clog dancer or human statue will perform. Kittens and puppies in cardboard boxes draw a steady crowd of admirers and adopters. Woodworkers, clothes designers, quilters, potters, silversmiths, glassworkers, oil painters, rug weavers, purse makers, photographers, decoy carvers, and many others offer wares. One man, who traveled extensively in China, paints exquisite T-shirts and dresses with oriental motifs; they often include poems in Chinese. Last year, when I told him about the great blue heron I'd seen at Six Mile Creek, he began painting dresses with great blue herons. I have several of his cotton dresses, and my favorite is a white one with pink and red peonies and a poem in black calligraphy which translates as:

In springtime in Jo-yang the peonies are like brocade—
All tones and colors, full, rich in spring's elegance.
The warm sun and gentle wind perfect spring.
My garden is filled with fragrance, wealth, and honor.

A dozen restaurants offer cuisine from macrobiotic to Thai. Organic meat, eggs, and goat cheese are sold. A woman who raises sheep and dyes the wool with luscious pigments is always knitting. For ten years I've never seen her *not* knitting.

Then there are the growers. One man, who raises 10,000 garlics and wears a long, straggly gray ponytail, looks a little like a garlic himself. Two older brothers sell black walnuts and homemade jams. Larger farms offer their weekly crop. One Saturday it may be strawberries and cherries, another week basketfuls of peaches, another apricots. People who sell fresh mesclun and herbs usually jar their own homemade salad dressings and jams, too. One woman grows mushrooms of many colors and varieties. Another woman sells edible flowers. Sampling is encouraged at most stalls. Nasturtium tastes peppery, tuberous begonia a little sour and lemony, pansy rather bland, bee balm a little minty. One might also adorn a salad with chives, English daisy, stock, or hibiscus. (But beware munching on flowers bought as houseplants, because they most likely were sprayed with insecticide.) Someone usually sells crystals, aromatherapy, fairy gnomes, or spirit aids. The man who sells honey in straws and dispenser bears also sells beeswax candles. Some growers offer cut flowers, which means the mingling crowd is full of people carrying bouquets.

Today, perhaps by previous agreement, a number of people have brought their exotic pets. I know of an agriculture professor who has a camel in his backyard, but I never see him riding or parading it around. That in itself is unusual for my hometown, a latter-day hippie community that seems at times to consist almost entirely of therapists and artists, which means that almost any form of eccentricity is taken in stride. I didn't realize how many people want to share their lives with unusual pets, or the degree to which they resemble their animals. Did they stare into a pool of water, see their features transformed in the wobbling mirror, and think a monkey or turtle stared back? Were they drawn in a totemic way, searching for a spirit guide, to the animal that best expressed them? Or, in time, did they simply begin to share expressions with their pet, as longtime spouses do with each other? Does the man who chose a snake as a companion feel it symbolizes an important aspect of himself? Because it separates him from "normal" society? Because it erases the civilizing words being silently uttered every time he puts on a suit? Why does a middle-aged woman with a Betty Crocker smile rejoice as her pet mouse runs across her shoulders? Is the woman of ambiguous sexuality, wearing the cutaway jacket and holding a large iguana with her right hand, using her left to restrain the invisible iguana in her brain? Or is she just thinking? Standing with the iguana in one hand, and the other hand on her forehead, she looks like she's in touch with the reptilian brain, the oldest, deepest, most primitive part of the mind. The analyst who is walking the twin bulldogs of his psyche—what am I to assume about his hounded home life? Has he noticed that the dogs look like swollen testicles? These animals don't strike a cute and cuddly chord in all of us, yet they obviously delight their owners, who move comfortably with them, cheek to cheek, heart to heart, side by side, or holding them aloft with dignity and affection, as if they were embracing a child or an intimate friend, with whom, at any moment, they might break into song.

As the exotic animals mix with people holding bouquets aloft like torches, the farmer's market bustles with the normal stir of a grand bazaar in, say, Istanbul, which I was lucky enough to visit with my mother when I was sixteen. I wonder what became of the sequined, curly-toed slippers we bought. They'd look at home here with blue jeans and a mint-green T-shirt silk-screened with frolicking otters. Soon I find my supposed reason for coming—a stall selling double-petaled rose petunias and hanging baskets artfully filled with mixed sun or shade flowers.

The homemade angel food cake sold by the man in the booth across the aisle will taste heavenly with some of the wild strawberries that nestle like small organs in my peppermint patch. I used to buy heavily budded plants from an elderly man who had a secret recipe for persuading his roses to bloom. Each week I would return home with another rosebush decked out like a Christmas tree covered in flowers instead of ornaments. He was a quiet man with a pleasant smile, who traveled quite a distance to market. I haven't seen him for several seasons now, but I hope he is well and still growing his contented roses.

23

At last it's time to remove the leaf dollies and leaf tresses and leaf mats and leaf thatches and leaf strangles and leaf coats and leaf shade-makers and leaf rain-stealers and all the other leaf annoyers clogging up the beds since May. I know the spring bulbs need to gather sun and rain and continue feeding if I want enthusiastic blooms next April, but about this time of year I lose my patience with them and even perceive the beds as suffocating under their weight. They *do* shade other flowers, which must be carefully guided around or through them. They *do* compete with young perennials desperate for a little generosity of sun and water. Bulbs need to stoke up for the next season, and in some beds I resisted leaf dollies, instead planting tall camouflage plants (mainly hostas, phlox, and daylilies) to rise among the faded kelplike narcissus leaves. I'm ambivalent about which plan works better.

At the end of the day, it feels good in some unnameable part of the psyche to inspect the garden, whose beds now show airiness and—dare I say it—even a little order among their extravagances. I'm not the sort of gardener who plants small islands of flowers, neatly separated by mulch or rocks. I think of that as zoo gardening. Not too much of anything, and everything in its place. I get twitchy just thinking about it. Those people probably have tidy desks, too. I like profusion. Although I admire gardens in which a single kind of plant mounds in great undulations of color, I'm not able to temper my own garden in that way. If eight-foot-tall wild mustard or meadow rue crops up in the middle of a daylily bed, I'm too surprised to remove it. Anyway, purple meadow rue, and mustard whose color is distinctive enough to bear its name, are also lovely plants which ride the breezes like gliders. If the blue lobelias have driven out the reds because they're competitive and always the victors, well, what can I say, but to the victor go the soils. Given that any garden is a willful act, I try not to intrude too much. The more the merrier is my motto, let my beds be an Ellis Island of natives and immigrants whose cultures blend into a beautiful mix. But the road to excess leads to the castle of indolence, and hodgepodge beds take a lot of upkeep.

I swear I can hear the freer breathing of the petite Stella de Oro daylilies as they lift their heads above the waves of sundrops. The Rockets have room to launch towers of yellow flowers in slow motion from the bottom to the top. Clumps of tall daisies are growing into colossal open-air bouquets I've encircled with peony hoops. The purple butterfly bushes have room to flex long floaty arms. The short, colorful zinnias have leapt out of the shadows. The self-seeding rose campion, which has spread itself all around the garden, finds room at last for its silvery stems and magenta flowers. The Victorians loved rose campion, and the mother plant from which mine came is about eighty years old, so it may well have been planted in Victorian days. How do I know its age? My friend Jeanne, a knowledgeable and delightful novelist, found a small plant overgrown by weeds when she moved into her historic house years ago. Thanks to Jeanne's generosity, and how freely the plant sheds its seed, most of the rose campions in town can probably be traced back to that mother plant. The roses are pausing between flushes, the five-foot-tall grasses are starting to raise their flags, and if I'm not careful I might start to lament the signs of fall, already present in the beds of newborns.

Some of the perennials are confusing to deadhead. The *Well-Tended Perennial Garden,* my Bible for such things, advises deadheading speedwell and verbascum. But with those plants there's only a subtle difference between seed heads and flower buds. The seedpods are a bit rounder and harder. The flower buds tend to point up a little more. Both have tiny hairs. Both states of the flowers—so different in purpose—can look identical. If you open a seed head you may find what's called "cheese" (because of the wedge shapes).

It's also time to deadhead the penstemon and campanula. Before deadheading the miniature hollyhock and the stachys, I sprinkle their seeds around the base of the plants they came from, hoping for a mob. A raggedly chomped stem catches my eye and then another and another. Oh, no! The deer have eaten all the new growth on the spring-flowering shrubs. I didn't know they ate sandcherry leaves. Well, that decides the height of those bushes anyway—they'll be low and compact. It also means I'll need to wrap them in bird netting or chicken wire for winter. Some days, it's so dry in the forest that trees begin wilting. Then the deer are happy to find succulent new growth in my watered garden.

I've added a curious new vase to my collection: a wooden hand with flexible fingers, which I thought it might be fun to play with. Now it sits on a near shelf, holding one red rose, the stem of which ends in a florist's tube of water. It looks pure surrealist, and I'm thinking of going back for a wooden foot so that I can arrange roses between the toes.

24

When Zen Buddhists create "dry" gardens, they often use only boulders, sand or gravel, and space. But there is a method to the arrangement, and the sand or gravel, raked into a wave pattern, gives a sense of flowing water. Another vital ingredient is weather, because freedom and serenity require adapting yourself to "forms which have the significance of cosmic laws." If wind or rain disturbs the carefully raked gravel, then you re-create the same pattern, meditating as you rake.

I don't think Zen and gardening go together easily in the West. My own brain seizes up at the thought. Finding a depth of moment isn't the challenge, because any garden contains uncountable moments of high drama or riveting calm. Gardeners live in the moment, but we also live in the future, and the past is always clearly in mind, too. Each flower has a history, a tale of struggle, perhaps, or disease. Each flower is planted with hope and expectation. A visitor admiring the garden may appreciate how it exists on that day in that hour, but to the resident gardener, who has raised the garden through many seasons, every moment is threaded with memories. Moments float against the backdrop of an imagined future, in which the garden looks even lovelier, if that's possible, or differently lovely, as other plants come to bloom.

Then there is the matter of lack of intention, which Zen requires. The garden should not seem beautifully arranged, nor should the gardener strive to create a garden that *doesn't* seem beautifully arranged. Instead one should be fully absorbed by a garden, not as the individual who created it, but as though it created itself in harmony with cosmic laws, drawing one's inner world and the outer world into balance.

Zen flower arranging, which began as a religious ceremony steeped in meditation and rigorous mental discipline, was esteemed by warriors, priests, and housewives alike. "Self-immersion in union with the flowers," Gustie L. Herrigel writes *in Zen and the Art of Flower Arranging* (1958), "breathes the very spirit of the Samurai, and the gravity of final, irrevocable decisions." One of its essential truths is the "principle of three," in which a person exists between heaven and earth, in communion with a plant as with the whole universe, becoming a channel for the spiritual as well as the earthly. Classical Zen flower arrangements include those three branches in ingenious ways. For me, arranging flowers is neither a religion nor a pastime; it's the quiet prelude to each summer day. I don't know if Gustie Herrigel is right, that "correct handling of flowers refines the personality," but I find it soothing to stroll among the plants, choosing this one or that for an arrangement. *Arrangement* in the other sense, as well, accepting the unknowns of the new day.

To still the moment, I repeat what I suppose is a mantra as I regard the garden on these strolls. "Beautiful, beautiful, beautiful," I repeat under my breath, as much a compliment as an observation. I smile and breathe deeply as I say it. I feel awed by the sense-tingling beauty of such life-forms so different from us. Simply beholding them treats our senses, and I am grateful. I don't expect understanding or response from the plants. I offer them my goodwill anyway, and the simple intransitive gratitude of "Beautiful, beautful, beautiful." Nor is it offered as a general hallelujah to life on earth, but to the individual plants, as praise, though they're not expected to hear or respond. Perhaps it is a little like hunters apologizing to their prey for killing them. I thank the plants for nourishing me. They do not mean to be beautiful, they cannot help themselves. At that point, I may enter a zone of transcendence, in which I marvel at all the accidents of fate, since the beginning of life on Earth, that led to my genes being created, and my standing in this particular garden with a contemplative and inquiring mind. I've been reading recently about how reflection evolved. What a fascinating solution to the rigors of survival. Human beings are an unlikely predicament for matter to get itself into. How amazing that a few basic ingredients—the same ones that form the mountains, plants, and rivers—when arranged differently and stressed, could result in *us*. More and more of late, I find myself standing outside of life, with a sense of the human saga laid out before me. It is a private vision, balanced between youth and old age, a vision in which I understand how caught up in striving we humans get, and a little of why, and how difficult it is even to recognize, since it feels integral to our nature, and is. But I find it interesting that, according to many religions, life began and ends in a garden. Creating an earthly paradise connects the two and offers a timelessness drenched in sensual pleasure. Striving doesn't necessarily stop in a garden. Quite the opposite—gardeners compete with weather, soil, neighbors, bugs, rodents, and common sense. But a garden can offer a tunnel through time, a sanctuary in the old-fashioned sense of the word, a sacred place where one is safe from human laws. I'm thinking of the laws we impose on ourselves, as well as society's, the family's, and then that something dimly lit and harder to fathom: instincts ingrained so deeply they feel like absolutes. For me, all those laws stop at the garden gate, and I can spend a small eternity with a rose.

25

I don't mind Japanese beetles having sex on the roses, I just wish they wouldn't eat at the same time. There are few things guaranteed to stop a rose lover's heart faster than finding a beautiful blossom being devoured by six copulating beetles. I try to pick the roses the second they're ready, to save them from the beetles, but it's not always possible. Today, on a tightly wrapped pink rosebud I find a beetle which has tunneled halfway through, gnawing ragged holes. I can tell where other Japanese beetles have been:

they perforate the leaves, turning them into fine lace and leaving tiny balls of black excrement on the petals.

I can't afford to miss a day's patrol. For some reason the beetles seem to prefer pink roses of a certain shade and shape. Occasionally I find one on a red, but rarely. It doesn't have to do with strength of scent—the most powerfully scented aren't necessarily their favorites. The Japanese beetle traps have a strong odor, but it must not be odor alone that attracts the beetles, or perhaps even what we call odor. Anyway, I often find a cluster of beetles having an orgy on one pink blossom, cutting chaotic paper dolls from its petals. I'm not sure why they also eat the leaves, but they do so in a characteristic way, leaving a lacework pattern behind. Why don't they eat whole sections of leaves? Why do they leave the outlines? To stand on? Too tough? They also love yellow flowers, and it's worth planting sundrops among roses to decoy the beetles away. When a beetle has begun eating a rose petal, the next day other beetles will home in on the site. Sometimes I leave a few sacrificial roses, partly eaten, and hope other beetles will attack them and leave new rosebuds alone. What attracts them? Does the beetle give off an "I found something good to eat" smell? Or does the plant give off a damaged-plant signal, less heat or electrical energy? Bugs are more attracted to the weak and damaged plants just as, on the Serengeti, lions attack the weak and old of a herd.

Like humans and most other life-forms, beetles are mainly alike but individuals differ, and isolated populations develop their own dialect, customs, and tastes. What works in my yard may not work in yours, but when it comes to deer or Japanese beetles, my best advice is try everything organic and keep your sense of humor. If possible, concentrate on their beauty. When curiosity glows, the fate of the roses doesn't look as dark.

Nothing on the east side of the house tempts the beetles, not the daylilies, Rockets, astilbes, or hostas. Many gardeners think of Japanese beetles as tiny Grim Reapers, symbols of death—the garden's and one's own. To me, the beauty of the Japanese beetles is what's disturbing. I brush them into a jug of soapy water, where they suffocate, but even in death their carapaces are exquisitely rainbow-colored. To be so beautiful and deadly, to be so beautiful and leave beauty slaughtered in your wake. . . . They're such pretty monsters. Yesterday there were few Japanese beetles around because it was windy. They don't like to climb and fly in rain or wind. But today is still, dry, sunny, and they are out in force. For a few years I hung traps containing rose scent and pheromone, but found it just attracted more beetles. At least one garden writer (who should best remain nameless) suggests you sneak into a neighbor's yard and hang traps there! Last year I applied milky spore to the lawn, where the grubs overwinter, and have reduced the plague considerably. It helps if you dose the lawn in consecutive years. Few bugs disturb gardeners quite as much as Japanese beetles, whose ancestors arrived long ago as stowaways. My severally gifted friend Michael Rosen wrote a grand poem about them:

But what of creatures we cannot justify?
Mornings, armed with a bucket of water and soap,
you comb the rose beds, plucking what seems to be

the sunlight's glint upon the leaves (if only
it were something so beneficent or brief),
but is, instead, Popilla japonica,

a nearly hundred-year-old accident,
buried in the burst heart of a rare blossom,
or flensing a leaf to its veinous skeleton.

Hard as it is, you must train yourself,
just as you have trained each hybrid rose
and still forbear its yearly disappointments

(black spot, mildew, winter kill),
that every drowning is not a victory,
either for roses or for humankind.

In the small rose garden, red Fairy roses are in spectacular bloom, clusters of pinkish-red flowers with a white throat. Chicago Peace (named in hope during World War II) has opened up two new blossoms. Red-and-white-striped Scentimental has begun to bloom again. The open, frilly-cuffed multicolored Playboy is covered in bloom. Pink Juliet has opened flat, petite apricot flowers, and Cardinal's Song has two large, fat red blossoms. No Japanese beetles in this garden!

The columbine seedpods have cracked open like tiny five-pointed starbursts. The snap-snaps (or wishweed) are so full of plump seedpods, I wonder if I have enough wishes for all of them. Forget the three wishes of fable; today I have fifty wishes, enough to share with friends.

Tall purple-and-white-striped phlox is starting to bloom. The ornamental lilies and daylilies are heavily budded, with many in bloom, and a long sprawl of colorful lilies at the foot of the yard looks like a plaza of international flags. Now is the time to strut and secure the dinner-plate hibiscuses, too, because they grow so topply and head-heavy that they stagger and fall when they open.

My garden rounds each day begin with a red wagon. I put in it a small cardboard box in which I stand two plastic containers—one with soap and water for the beetles, one with water to hold freshly cut flowers. Next I add some metal struts, a roll of coarse green string, my deadheading and flower-clipping scissors, a pair of cotton work gloves, tools, and my garden journal (inside a plastic bag, to protect it from spills). Essentially, I take the gardening shed with me as I migrate from bed to bed.

In mid-summer the garden is like a resort town that relies on its annual visitors for color and intrigue. Dahlias and zinnias, sunflowers, marigolds, and petunias, geraniums and gladioli will fill in the empty spaces through fall. When they finish, it will be winter, a season with its own stark beauty, when the garden reveals its bones, until spring when the garden's native residents return once more.

I awoke to the sound of steady rain—hard, small drops I could hear picking at the windows—and knew the garden would be grateful. Soon the sky cleared, and now a hot sun is blazing that will shoot the day's temperature to 90 degrees before it's through. Even lying still in the shade I begin to overheat, but two sweat bees are sipping from my arm so gently I haven't the heart to dislodge them. A handsome red dragonfly, whose body pikes like a diver in midair, sits on the slate, then it spreads its double wings like a biplane and soars. High puffy clouds are maneuvering at speed; the winds aloft must be fast. The slug traps are working, but it's not much fun to scoop slugs out with a slotted spoon and toss them into the woods. Funny how we learn to recoil from squishy, slimy things.

It's time for daylilies, and the yard suddenly has height and architecture as the brilliant flowers wave in the breeze like regatta flags. The daylily hedgerow at the brink of the woods is blooming with dozens of shapes, smells, and colors. I planted Franz Hals daylilies today, each one a tall duet of yellow and ochre. The six petals alternate: yellow, orange with a thin yellow stripe down the middle, then yellow again. Many of the lilies are startling because of their contrasts: brilliant scarlet with golden throats, yellow petals branded with a garnet six-pointed star. Pandora's Box, one of my favorites, is a small cream-petaled lily with a magenta Rorschach in its throat. There are lilies with lime-green throats, black-red lilies that look like congealed blood. I've also planted some tall, old-fashioned, wild orange daylilies of the sort one sees along the roadside, and two spectacular blazing red-and-yellow Open Hearths. Each year I add more late-blooming September Golds. They *are* golden, not yellow, with big stiff cups. Everything is blooming so early this year that I'm sure the September Gold will bloom in August. Then what? The gardener's panic.

The lavender is going to seed, its purple flowers growing into shaggy purple rags, but the blue-green leaves still smell pungently spirit-lifting. Nearby, curly mint is now three feet tall. I've always meant to plant a path of herbs to be walked through.

When the sundrops have finished, and the daylilies are in full color, and the phlox rocks brilliantly in the sun, a sad time of dread and premonition begins to taint my mood. At the nurseries, a few straggly unsellable roses remain. Sales offer flowers that were premium only a month ago. Late-blooming daylilies foretell the fate of the garden. Most distressing is the corn in the fields. For weeks, I have been watching it shoot up. I swear you can see it grow. Corn is an ancient plant. People have told time by the corn clock for millennia.

"No, don't let the summer end," I beg Paul magically, childishly.

"Summer doesn't end here until November," he says, hoping for a long swimming season.

I remember how Emile Zola described the perfect bell jar of summer, in which the garden was nearly animal, a being with whom one shared a secret paradise. "Under that heat haze the great garden lived like a happy beast," he wrote, "released from the world, far from everything, freed from everything."

26

By August, it's hard to keep flowers blooming. I take an informal inventory and state of the garden nation. At last the hordes of Japanese beetles are retreating, the black spot is surrendering to pruning and sulfur baths, the roses have received their last feeding of the year and are responding with new red leaves, which signals another burst of buds. The red acts as a sunscreen for the delicate new leaves. Showy sedum has just begun to blush pink, white, and red. White star-flowered hostas open under the apple trees, beside tall, yellow-and-red-capped sneezeweed. The blue obedient plants, pink turtleheads, and towering red lobelias rock in the breeze. Red, yellow, and pink begonias continue to blaze like neon. But the waves of daylilies have finished, and I hate to see them go. I've ordered twenty very late bloomers, whose double fans I'll plant to enjoy next year. One is always investing in the garden—it is a growth fund. The gloriosa daisies in front of the bathroom window have quadrupled in number and soared to seven feet, creating a screen. I'll move them in November to the bed of yellow things: sunflowers, black-eyed Susans, euphorbias, yellow roses, perennial yellow foxglove, and yellow mums whose distinctive gray-green leaves are ribboned in silver. I've decided to rethink a front bed that's mainly subtly hued purple and yellow daylilies mixed with giant purple alliums. Lilies will thrive in part shade, so I'll move them, and devote the sunny bed to wildflowers: chicory, dame's rocket, Queen Anne's lace, bladder campions, teasels, milkweed. Although I love how milkweed pods burst with silky parachutes, I may have to resist them, since their underground root system spreads far and fast and they can choke a small garden.

It's been 94 degrees and dry for three days straight. In my experience, when the outside air rises close to body temperature, we tend to feel sick, as if our borders are melting and there's no way to tell where we stop and the day begins. The roses on the patio are shriveling. A good soaking revives the rest of the garden. The September Gold daylilies are thick with rich orange-gold flowers. The purple rose of Sharon is covered with red-hearted blossoms, and my beloved roses continue to bloom. The Abraham Darbys are starting all over again. Curly peppermint is a shag carpet of frilly lavender plumes. I pick a bunch and arrange them with yellow, red, pink, and orange roses.

A bowl of roses sits on the marble tabletop. Sunlight shines through their petals as if through stained glass. The Othello is luminous, a deep, saturated magenta that radiates color like heat. The apricot Polkas are delicate as tissue paper. Because light shines through the flared petals of the red roses, but not through the tightly wrapped buds, they appear to have halos around their heads. The red-and-white Scentimental isn't just striped; some of the white petals look splattered with red. The splatter pattern carries one's eye outward and gives a sense of explosion to each flower. Ruffled red-and-white explosions. This is the summer of the orange roses. Dozens of brilliant orange blooms of Lasting Peace, Singin' in the Rain, Abbaye de Cluny, Polka, and Tropicana. That doesn't include the pink-apricot roses, such as Sweet Juliet, Abraham Darby, and Colette.

New buds forming on the roses promise another glorious flush at season's end. The wren parents—we call them Jenny and Christopher—are shuttling back and forth to feed their nestlings from what seems like an endless fund of crickets and caterpillars. Either they have few nestlings or wren chicks are just low-key, because I hear only a few soft peepings when I walk by. Mind you, I can't get very close for long. When I went into that part of the garden this morning, the male wren scolded me so fiercely that I was momentarily offended. His loud chittering is a danger call, a warning of nearby threat, and frankly, I don't consider myself a threat to them, but rather an affectionate well-wisher. "Okay, okay!" I said, backing up and moving to another bed. When I was twenty feet away, the male sounded the "All clear!"—a loud version of his usual bubbling song. Because hallucinations stalk the glass, the wrens don't see me sitting in the bay window, at eye level with their birdhouse only a few yards away.

The parents run a proper shuttle service now, providing hundreds of mouthfuls each day. When their paths cross at the nest box, they do a fluff-and-quiver dance on nearby branches, to plight their troth. Each time they arrive with a small green tidbit, the babies squawk in unison. I can just imagine their urgent hunger.

Only two weeks ago, the female stayed inside the nest with the chicks and the male brought food. I don't know if she ate what he gave her, to keep her strength up while tending the young, or if she was there to macerate the food first and then feed it to the young. It's always easier if a parent predigests food a little for its offspring. We do the same thing by cutting up and cooking food. In fact, that's the most likely origin of French kissing. In the distant past, human moth-ers prechewed food for their infants, just as birds and other animals do, and passed the food by mouth and tongue into their baby's mouth. When infants grew up, they retained loving associations with deep kissing. In chic big-city restaurants, you can get what's called "macerated" greens, and whenever I see the word I'm reminded of all this and how we've never gotten over our yen for semidigested food, or for kissing.

More and more, though, people are eating fish or beef grilled on the outside and left raw within. "Too rare?" a waiter asked me one day, as I handed back a tuna steak for further cooking.

"I've seen animals die from worse wounds," I grumbled. But I did notice other diners devouring the dish with great relish. I don't think it's a coincidence that we prefer our food cooked to the temperature of freshly killed prey—at least on the outside—but I try not to disturb dinner companions with such thoughts. This goes back to the "Mind your own garden" rule. Recently, I saw a bumper sticker that said: "I don't eat anything with a face." That would be rude, too, when I dine with friendly carnivores.

Though mainly a vegetarian (or, if you prefer, someone in league with seitan), I've never grown a vegetable garden. I envy those who do, but in my yard that would mean taking a number behind such a long line of vegetable lovers—raccoons and squirrels and groundhogs, birds and insects—and demand constant vigilance with little reward. Also, I only eat organic fruits and vegetables, and organic farming takes a lot of labor. I bless all the kind souls who devote their lives to it.

Poisons would work better and faster on Japanese beetles, but knocking them into a pitcher of soapy water is healthier for us and the environment. Poisons would work better on black spot, but using sulfur is healthier. Poisons would work better on rose worms and aphids, but soapy water will work well enough. There's nothing like old-fashioned soap to banish most pests in the garden. I even spray an oily soap called Hinder to keep deer from sampling plants. I begin spraying it once a week in spring (or after a heavy rain), and as the summer unfolds, fewer deer dine here. They regard my yard as a place where food tastes soapy. If you're willing to poison yourself and the ecosystem to have a well-tamed garden, then what is the point of the garden? Certainly not nourishment.

Away for two days, I return home late at night. The next morning, I'm amazed to discover the garden continued growing in my absence. I carry with me an image of it as it was. As I settle in my bay window, I notice something strange: stillness, silence. No male wren singing from his favorite perches. No female wren rocketing through the magnolia branches. The nestlings have flown! The birdhouse looks fine and untampered with, thank heavens. Unlike the wrens that lived in the federalist birdhouse, this family has succeeded in fledging its young. I miss their customs and beautiful song, but I'm also happy for the chicks, which were always learning to leave.

Another surprise: a female hummingbird appears at eye level, then flies to a nearby fence, where she perches a long while. I know it's a female because she doesn't have a red throat, but is dusky brown with an iridescent green streak down her back. She heads for the orange trumpet vine and begins sipping nectar, then moves to a lone clematis blossom.

A small article on the front page of the local paper this morning reported that the Oxford English Dictionary would soon appear in a new edition—the first in seventy years. It will include 2,000 new words that have entered the language since the 1970s, expressions like "road rage" and "pot pie." "Snuck" will be acceptable as the past tense of "sneak." We're not usually aware of the language evolving under our feet, or tongue. Although I'm fascinated, I also find it a little disturbing, since it's a sign of time passing, life quickening away from me. Unfortunately, the Oxford English Dictionary doesn't include all those wonderful words in other languages for which we have no English equivalent, words we desperately need, such as the Tierra del Fuegan *mamihlapinatapei,* which means two people "looking into each other's eyes, each hoping that the other will initiate what both want to do but neither chooses to commence." Ah, yes, a sweet familiar anguish that might occur in a garden. Or the Russian word *ostranenie,* which is when an artist makes the familiar seem strange, so that it can be seen freshly. Or *aware,* the Japanese word for the special poignancy one feels while enjoying ephemeral beauty. Or the Indonesian phrase *holopis kuntul baris,* which summons extra strength for carrying heavy objects. A gardener experiences *aware* and *ostranenie* throughout the seasons, and who hasn't wished for a magic strengthening phrase?

Four years ago I planted a single obedient plant—tall gothic spikes of purple flowers—which multiplied fast, and I began to ignore it. Last year I moved some of its offspring to a shady border by the woods. Today the obedient plants give that border architectural lines of a European cityscape seen from afar: gothic spires on churches. I can't miss it, and it loves the moist shade, where it grows tall and straight, without pleading for sun in a gangly twist.

"It's happy there," I say in gardener's shorthand to Chrys one day. We both know how delicate the right balance can be. All living things have ideal environments for growth. Sometimes they have to be moved around quite a bit, and their needs analyzed more subtly, before they take root in a place well suited to most of their needs. Then they thrive. *Growth.* Only a word, it sounds like a simple process, but it is the combined destiny of so many processes.

This year, for the first time, I thought well ahead and put peony hoops, open fences, poles, and struts on many of the plants I knew would shoot up thick, tall, and floppy, spilling all over themselves and breaking from their own weight. The turtleheads and obedient plants are well supported. So are the dinner-plate hibiscus, still in bud, which will produce fifty huge heavy red or white flowers in September. By then, bracing them would be difficult. So now, while it is still manageable, I carefully tether their stalks. Some flowers do best when aided by restraint. Otherwise they may perish from their own weight. This is also a problem for some animals, like beached whales, which can die from their weight when out of the water. Heavy flowers on slender stalks are always at peril of breaking their necks. Today I put hoops around the base of the bushy grasses, too, because last year they draped their long beautiful plumes over the driveway. Driving through a gauntlet of plumes can be fun the first thirty or forty times, but then the plumes start getting battered and string does little to gather them. Guided growth is better. Giving them a smaller circle in which to maneuver forces them to grow straight and gives them a little metal to fall back on in rough weather. Soon the new cityscape of obedient plants will multiply and begin to invade the woods with purple spears. Then they'll be disobedient.

27

A chill in the morning air foretells the fall. Last night the cicadas piped loud as Scottish infantry, skirling their whereabouts and mood across the forest. During a television newscast, a reporter from Washington, D.C., standing in front of the White House, could barely be heard for the outcry of the cicadas. I am planting forty lobelias to spire and spread wherever yard meets woods. Good in the shade, with beautiful towers of orchid-like flowers, lobelias don't mind marshy soil or even clay as thick as pottery. But I won't plant the pinks and blues together. They fight, and the blues always win, since they're made from wilder, tougher stock. In Latin, the plant is called *Lobelia siphilitica,* because a head of Indian affairs sent some to Europe as a cure for syphilis. Unfortunately, it didn't work. As it turned out, lobelia was only one vital ingredient in the Indian tonic. Many garden words contain the social history of their use and the people who for one reason or another developed a special relationship with them. Consider the case of phlox, a tall garden perennial of many colors that produces large clusters of flowers enjoyed by deer and humans alike. Vita Sackville-West thought it smelled like a pigsty, but since I've never actually smelled a pigsty, I can't confirm that. To me, it smells gluey and is a miracle of magenta, red, and white in the summer and fall garden. The annual phlox, *Phlox drummondii,* is named after Thomas Drummond, who found it growing in Cuba and sent specimens back to Britain. His story is bizarre, but not at all atypical of botanical explorers. Here's how Diana Wells summarizes it in *100 Flowers and How They Got Their Names* (1997):

> Drummond was curator of the Belfast Botanical Garden and went to America in 1831 as an independent plant collector, exploring much of the Northwest by himself. He sent his guide away and spent one winter completely alone in a brush hut. He survived by chewing on an old deerskin when, because of snow blindness, he could not see to shoot game. He deterred grizzlies by rattling his specimen box at them, but this was less effective when he got between a mother and her cub and was nearly killed. He later survived a shipboard epidemic of cholera, nearly starved while wintering alone on Galveston Island, lost the use of his hand for two months, and suffered such severe boils that he was unable to lie down. In spite of all this, he applied for a grant of land in Texas, intending to bring his family over to America. In the meantime, he went to Cuba, but in 1835, he died there from urecorded causes. One of the last plants he sent home was the *Phlox drummondii,* which Sir Joseph Hooker of Kew named in his honor, to "serve as a frequent memento of its unfortunate discoverer."

Even if we don't remember him when we enjoy our willowy phlox, his namesake flower is so common in American gardens that, in a sense, he has homesteaded here after all.

Just as the Japanese beetles are disappearing, the boring caterpillars begin to teem. The caterpillars may be the toughest and most dangerous rose plague because they tunnel into the unopened buds and destroy them on the eve of blooming. In a real sense, *they nip them in the bud.*

"Oh rose thou art sick, " William Blake wrote. "The howling

worm in the storm has found thy secret lair." Nearly impossible to spot, the caterpillars are perfectly camouflaged in the same green as the leaves. If you're lucky you may see one curled up on a prominent glossy leaf, but more often they choose the angle where leaf meets stem and are very hard to spot. I pick them off by hand. Long ago I gave up being squeamish about touching the rose pests. This adds much extra time to my daily rose patrol, because I make two passes looking for rose worms: the first a general search of obvious leaves, the second more methodical, looking especially for leaves perforated by their telltale feeding. At least then I know worms *were* on the rosebush. But where are they now? The best way to spot them is from below a leaf, looking up as it's illuminated by the sun. Then one can see silhouettes coiled up like tiny cobras. In *A Midsummer Night's Dream,* Queen Titania sends her elves "to kill cankers in the musk-rose buds." I could use some of those elves. Today I found twenty worms, and I shudder to think how many I missed.

The tall grasses along the driveway, fluttering like hula dancers, are growing heavy and tall. Some dinner-plate hibiscus have begun to bloom—their flowers like blood-red elephant ears. Others are white and as big as old LPs. When they fall off they leave behind what looks like a stiff gold egg-cup crown, which also dries and drops. Carefully, I again brace the top-heavy hibiscus with hoops, stakes, and assorted stanchions. It's surprising how much time one spends in strutting things up, bolstering, angling. Many plants that thrive in shade—red cardinal flowers, for instance—need serious support, and as they grow, the supports must grow with them. Any flowers I forgot to cut back in spring now grow bushy and tall but have fragile stalks that break easily. Next year I'll see if I can train some of the young plants to grow through peony hoops and crop others early in the year so that they'll grow lower and bushier. But most of the strutting I do is to angle blossoms into better view. Garden as flower arrangement? Sometimes, yes.

A helper has taken one of the fallen hibiscus flowers, put it in a jar full of water, and added broken-up mint leaves. Earlier he was eating small slivers of echinacea removed directly from the flower head. "Does it heal?" he asks of the six tall boneheal plants I'm stationing as a backdrop at the foot of the yard in a transitional place where lawn meets woods. Soon the white clustered flowers of the boneheal will bloom and make a heavenly backdrop for the purples, yellows, and blues of speedwell, rudbeckia, and lobelia. Most often it's known as boneset, though not actually used for broken bones. Effective against the nineteenth-century scourge of breakbone fever, boneheal was classified as a fever-break useful in treating malaria, flu, and other illnesses. Apparently, it causes ample sweats in the patient, and sweating down a fever is traditional. This is a variety of boneheal called Chocolate because of its shiny cherry-brown leaves. Since boneset is related to Joe-Pye weed, which supposedly has the same medicinal qualities, I intermingle the plants. They seem to get on well, and it's an easier transition from nursery to yard if relatives surround the new plants. Joe-Pye weed was named after an Indian medicine man who used it to cure typhus, thus establishing his credentials in New England. I also brought home six bushy, chocolate-red snakeroot plants, heavily budded and ready to bloom into clusters of white flowers. Cousin to Joe-Pye weed and boneheal, snakeroot will fit in nicely beside them at the woods' edge.

The sweet cherry season has finished, alas, but the grape season is in full swing, and the small, brownish-red Reliance grapes are exquisite. No seeds. They're wonderful mixed with black friar plums, the first crop of Bartlett pears, and any of the dozen or so varieties of apple grown in the Finger Lakes orchards. My friend Sheila is swimming across Cayuga Lake at noon, as she and some pals do annually. Although the lake will be at its warmest temperature all year, it may only be in the 70s, which feels cold after an hour, even if you're in constant motion. But lake waters can be considerably warmer on one day than another. Most of 200-foot-deep Cayuga stays cold, since cold water falls and doesn't mix with the 10-foot layer of warm water on top. But a south wind will blow the warm top water north, leaving our south end of the lake colder. With a north wind the opposite is true, and, fortunately, a steady wind is blowing from the northeast today, promising the swimmers a warmer crossing.

I'll never get used to how fast things grow. Yesterday I deadheaded the zinnias and butterfly bushes, and this morning newly dried flowers await my shears. When did that happen? Overnight? The showy sedum called Autumn Joy has begun blushing pink across the top of its needle-cushion flowers. If broccoli florets were pink, they'd look like this. The huge stately grasses ride the wind, bowing low over the driveway, then rearing up with a flourish. In the small rose garden near the bay window, I find a deep red Cardinal's Song ready to bring in and, to my special delight, a Paris de Yves St. Laurent. I've been waiting for this unusual rose to flower again, because I love the tight pink fist at the heart of each flower as it opens. Most roses will show their sex organs as they fall open in the last wanton stages of bloom, but Paris de Yves St. Laurent keeps its nether parts hidden behind a pink palm. There is something unavailable about its final mysteries, something of the Persian garden of delight.

As I do my garden rounds, I can feel the weight of an oatmeal raisin cookie in my pocket, so I pause beneath an apple tree to chew a few mouthfuls while surveying the shade garden. I see the deer have paused here, too, and chomped off hosta flowers. The slugs stopped here to eat holes in the hosta leaves. I wonder what makes the hostas so tasty to deer and slugs. The deer cropped off most of the astilbe's feathery flowers, too. Busy night for them. They left brilliant red impatiens and orange-and-yellow tickweed. Don't they care for them? Or were they too lazy to trot the few yards back to where they grow, flanked by pink turtleheads and rhododendrons?

The female hummingbird has started visiting the feeder, I presume because some of her favorite nectar flowers have finished for the season. I've named her Greensleeves, because of the green flash on her back. Her mannerisms are different from Ruby's. She's nervous, bouncy, skittish. This may be because she knows she's in his territory. Anyway, they took turns at first. Ruby preferred the nectar hole on his right, farthest from the door. She drank from the one on the far left. They never dine together, but clearly are aware of each other. They may even see one another in ultraviolet, as they do the nectar guides on flowers. Once, only once, I saw a baby hummingbird at the feeder. Dusky and smaller than the other two, with a short bill, it appeared suddenly in one fluent arc, sipped briefly from all three holes, and flew away in the direction the female had gone. I named him Gizmo. Hummers only lay two eggs per season, so this may well be her only chick. Do other hummingbirds visit my garden? I grow many flowers they relish. For example, under the apple trees, the hostas offer white bell-shaped flowers on long flexible stems. Some of those large hostas smell heavily of lily. You have to plunge your nose right into the flower to enjoy its powerful aroma, but that's the best way to smell most flowers. Four butterfly bushes—lavender, white, yellow, and dark purple—could keep a bird busy for some time. Between the flowers and the feeder, there's plenty of nectar to go round. But males are extremely territorial. Yesterday, male and female bumped into each other at the feeder, and he chased her away at sword point.

A sphinx moth, perfect as a botanical drawing, hovers nearby while I deadhead a large butterfly bush filled with tiny red-eyed purple florets. Sphinx moths move a lot like hummingbirds and are often confused with them, in part because they favor the same flowers. But a sphinx moth's wings move more slowly than a hummingbird's. You can see the wings in motion, looking a little like the brim of a bridesmaid's hat, or a blur of stiff caramel, or a photograph of Saturn's rings. It's a trompe-l'oeil. The remembrance of motion netted by the eyes is dragged from moment to moment as a curving sheen. Although I'm elephantine compared to it, the sphinx moth doesn't seem afraid, as it dips into one floret after another, sipping nectar and going about its business as I go about mine.

28

"A Guest in the Garden"

When I'm not gardening, I love to perch in my bay window, surrounded by greenery, and watch the ocean of weather and the ways of humans. It reminds me of lazing in a hammock on the prow of a schooner, with the sky above and the water below. Remembering the dolphins that rode the bow wave of that schooner, I begin to feel a familiar restlessness. There's so much wilderness yet to explore, including the oceans and space, and something tantalizing about the edge of things, the back of beyond. Earth is a garden in space, full of blooming life. Do other planetary gardens exist? I imagine so. Our science fiction scares us with alien monsters and suffocating climates, but not long ago similar myths abounded about Earth's unexplored regions. People set out on pilgrimages into the wilds anyway.

I've been reading the sensuous notebooks of John Muir, the wandering mystic who founded our national parks. I hadn't realized that Muir was a gifted inventor first—who refused to file patents because he believed discoveries belonged to all—and turned his back on considerable fame and fortune to seek a saint's life in the wilderness. He was part Zen, part zealot, but wholly American in his idealism and environmentalism. Not only our national parks but our attitude about America's landscape and "wildness" come from the obsessively kept journals of John Muir. At the end of Emerson's life, he listed Muir among the handful of truly great men he had known. I wish he were here this afternoon, because I've been thinking about his passion for flowers, which he describes as "plant people . . . standing preaching by the wayside," and also about his epic pilgrimages, not merely to witness but to become the soul of the wilderness.

It's an old tendency of humans to leave home and strike out across a frontier that beckons as a zone of magic, mysticism, inspiration, and holy conversion. When we are at loose ends emotionally, we tend to set out on a symbolic journey into unfamiliar territory, where newly aroused senses allow us to feel vigilant and reborn. In part this is based on the intuition that to change one's self one must relinquish all that is known and habitual, cast off from the shore of one's home and the endearing familiarity of everyday life, whose moods and manners one comes to know like an old friend. It is one of the more serviceable ironies of literature that we do not always travel to escape our circumstances but to find ourselves. Why must we do that in a foreign place, having become foreign to our past?

When Moses led his flock into the wilderness, it was not just to separate them from their oppression. Taskmasters may die or be left behind, but oppression travels light as a virus. Outrunning the psychic burden of oppression may indeed take four decades. In religious cults of all sorts, priests or leaders separate their followers from society, strip them free of as many old habits as possible, and guide them across the wilderness. The wilderness may be an actual frontier fraught with danger, or it may be a wilderness of doubt. Rebirth is the usual metaphor, one that occurs in so many rites and creeds throughout the world, even among people who claim to be atheists, that it must be a powerful instinct. On a fundamental level, we understand the biological script we follow. We know how parenting sculpts one's character and heart. We know that changing a reluctant and sedentary self will take equal force. The monument must become liquid again, be broken down to its elements and re-formed into another shape. The metaphor we choose is rebirth, the guide through that process we identify as a parent, usually a father. Sometimes we identify Earth itself as a mother. In some stories, including life stories, the pilgrimage is recognized as such. In others, it happens more picaresquely. The element of renewal can be dark or light. Reinventing oneself has a powerful appeal and works equally well for both saint and con man.

Up to now, I've been presupposing that one journeys to find oneself or something. Sometimes people travel just to lose themselves. Lose their past, with all the expectations it held. Lose the suffocating laws of society. Lose the infantalizing of organized religion, which too often encourages people to "should" all over themselves, as Albert Ellis once said. Lose the inner compass—moral, family, whatever—that kept pointing them toward the absolute north of some conviction. Lose all the clutter of relationships. Lose the pursuit of happiness or success in the regard of others. Sometimes one travels to lose the possibility of loss. Pilgrimages are on my mind, because I've made so many real or imaginary ones in my life, and I like to read the journals of other naturalists making their own. I doubt I'll ever get used to this quirk of mind: that one can be sitting quite still in a backyard garden, reading, barely feeling the sharp edge and nearly weightless pull of paper as one turns a page, surrounded by local sights and sounds, while vividly picturing oneself in a landscape a continent away. How we evolved a mind that can do many things at once and even imagine itself being otherwise, I will never know, but I'm grateful. Of course, this also means we can imagine states of perfection we can't achieve, worry over misfortunes that don't happen, or keep chasing a better life around the bend.

John Muir tells us little of his motives, except to hint at a childhood ruled by a fanatic father. The temptation is to picture him as a roaming mystic, mesmerized by the landscape, ever drawn to the horizon. His was a mindfulness taken to its limit, where it becomes ecstatic freedom from self. Reading his journals, one doesn't learn about his inner life—I mean the ongoing conversation one conducts with oneself about family, friends, insecurities, important choices, and so on. No doubt he had them, but they rarely make it onto his pages. Only his ecstasy, transcendence, and awe. The journals exist not so much to communicate something as to help formulate his joy, to make it more enjoyable to him by giving it words. "After dark, when the camp was at rest," he writes,

> I groped my way back to the altar boulder and passed the night on it,—above the water beneath the leaves and stars,—everything still more impressive than by day, the fall seen dimly white, singing Nature's old love song with solemn enthusiasm, while the stars peering through the leaf-roof seemed to join in the white water's song. Precious night, precious day to abide in me forever.

He does lament, though, how inadequate language is for such a transcendent state "which has no word symbol on earth." Muir writes poetically and even creates new words at times, for example, "the mist-lakes." There is little difference between that and the chimpanzee taught American Sign Language who, when she saw a duck for the first time, signed "water bird."

Anyway, we never hear of his physical ailments or even sensory responses—except to the extent that they're revealed in his observations of nature. Those can be quirky and astute. Here he is describing "certain spiders . . . which, in case of alarm, caused, for example, by a bird alighting on the bush their webs are spread upon, immediately bounce themselves up and down on their elastic threads so rapidly that only a blur is visible." Living outside of himself, he wishes to be a shard of glass, as he puts it, a pane through which nature becomes visible. Not a lens or a window, but a shard, something ragged and haphazard with rough, dangerous edges, that nonetheless reveals the world's beauties. In this, he sounds like Emerson, who wished to be an eye traveling through nature, observing and recording. Unlike Emerson or Thoreau, though, Muir was not afraid to gush. His sense of thrill reached delirious heights. One day in June he writes, in a biblical plural that includes all of humankind,

> We are now in the mountains and they are in us kindling enthusiasm, making every nerve quiver, filling every pore and cell of us. Our flesh-and-bone tabernacle seems transparent as glass to the beauty about us . . . thrilling with the air and trees, streams and rocks, in the waves and sun,—a part of all nature neither old nor

> young, sick nor well, but immortal. . . . I can hardly conceive of any bodily condition dependent on food or breath. . . . How glorious a conversion, so complete and wholesome. . . . In this newness of life we seem to have been so always.

Proustian in his alertness to the look and feel of his surroundings, he was jubilant about the simple being of natural things. Had he read Darwin, he might have liked the latter's word "panmixis," which Darwin invented to capture how the everythingness of everything reminds one of the everythingness of everything else.

On at least one occasion he climbed up a tree in a thunderstorm, wishing to be ravaged by the wind and rain, clinging to the erect, swaying trunk as his pulse sprinted and he cried out in ecstasy. Turner, the British painter of tumultuous storms at sea, used to tie himself to the mast of a ship and sail into the middle of a raging storm so that he could thrill to its fury and fluctuating colors. But Muir has more in common with John Donne, who writes in one of his loveliest sonnets: "Batter my heart, three-personed god . . . unless you ravage me." To offer oneself totally, to the ends of one's being, body and soul, includes the vigorous and erotic body as well as the delicately transcendent soul. Anything less would be hypocritical or a failure of nerve. It's hard for people to understand that sometimes. We separate the sacred from the profane, and even set them in opposition to one another. In many religions—both so-called pagan and so-called traditional ones—intense sexuality can be holy. Christianity isn't one of them, though, and when the church talks of Christ's "passion" it means something decidedly nonerotic. Hence nuns "marrying Christ" or churchmen taking a vow of chastity. Donne's relationship with God was intimate, personal, and profound. So was his relationship with women. Should his relationship with God arouse his emotions less than his relationship with women? In his poems, the two relationships blur, and when he speaks of God he uses forceful sexual imagery—minus the flirtation, playfulness, or uncertainty. He offers up his complete passion, the total sum of his most powerful emotions, and thus his relationship with God combines the rare harmony of casual and formal love, delicacy and ferocity. It suggests a love that embraces all forms of human love.

In his passion for the natural, Muir doesn't realize that civilization is natural, too, merely an interesting stage in the life of hominids. Is culture what one creates when one has lost nature? Muir was a man who lived by that axiom. To him, culture and nature do not intrude on each other. Furthermore, they must be kept separate because humans are a contaminating species. In his vision we dirty the planet and ourselves. "Man seems to be the only animal whose food soils him," Muir writes, "making necessary much washing and shield-like bibs and napkins. Moles living in the earth and eating slimy worms are yet as clean as seals or fishes, whose lives are one perpetual wash." He believed in keeping nature pristine and wild. If he were sitting with me in my garden, we might end up arguing about this view: his emphasis on our being *among* but not *of* nature. But on the poetry of everyday life, in which nature is a pure flame that singes with its beauty, we would agree.

Though not as poetic as Donne, nor as introspective, nor for that matter as happily civilized, Muir is part of the incendiary tradition to which John Donne and Gerard Manley Hopkins and Walt Whitman belong. For Muir, as for the others I just mentioned, sensuality combines naturally with reverence. Indeed, reverence without such sensuality would feel fake or halfhearted. Thus Muir doesn't just praise the lakes and streams and the water spiders that stride across them. He praises the smell and taste of the air—life's processes, including the sexual. When asked my religion, I usually tell people I'm an earth-ecstatic. But Muir was more devoutly so than I could ever hope to be. That he believed in a thinking, active God was almost beside the point. He lived in the church of the pines and worshipped creation at the level of twig and mole. He praised each one of "the plant saints that all must love and be made so much purer by every time it is seen." He may actually have prayed from time to time, but his notebooks are themselves a form of prayer, and although he might not have said so, I believe they would have fully satisfied his need for worship. They record a certain liturgy of sky and cloud, a creed of natural laws, and many moments of illumination. But he records them intimately and in exquisite detail, with an urgency bordering on panic, the way one records the unique shapes and shadings on the body of a lover. Here he is responding to a large cedar, a tree described as being strong and erect as a pillar, with yellow-green foliage and cinnamon-colored bark:

> I feel strangely attracted to this tree. The brown close-grained wood, as well as the small scale-like leaves, is fragrant, and the flat over-lapping plumes make fine beds, and must shed the rain well. It would be delightful to be storm-bound beneath one of these noble, hospitable, inviting old trees, its broad sheltering arms bent down like a tent, incense rising from the fire made from its dry fallen branches, and a hearty wind chanting overhead.

There was a time, before birth, when the outside world was inseparable from oneself, when world and self were united. A time of at-one-ment, lost forever at birth. Is it still remembered at a preconscious level in the body's awareness of self, and in our inevitably frustrated search for what we call peace? A return to that state of blended self and world can only occur in death, as Dylan Thomas wrote so movingly in his sonnet "When All My Five and Country Senses See." Perhaps that is why we find the two states so compelling. Why should death be compelling? Organisms wish to continue and reproduce. Is death compelling as a relief from the stresses and rigors of consciousness? Or is it a voluptuous pull toward the realm of perfect at-one-ment that we remember in our cells? For short periods, we can imagine that idyllic state of wholeness.

Opening up to the wilderness is an act of pure hope; like a feral child, one absorbs every experience and sensation. I find exploring my garden world as replenishing as an expedition, though I've been on some lulus. The ultimate nature mystic, Muir would undoubtedly have insights to share about eco-spirituality, and he'd be a welcome guest. There's nothing like wide thoughts in a small garden. I also think he would enjoy hiking the Finger Lakes trail that winds through the forests of this region.

Was Muir also stirred by lust? Of course. In time he married and raised a family. And there was his curious affair with a British author of steamy, bodice-ripper romances. Evidently, Muir lost himself in her arms at times, and she pursued him through the thickets of Wyoming and social convention. From all accounts, she was her own weather system. And in the end she returned to England and he to the undulations of the land and the robust, but impersonal, demands of seasonal time.

29

Time is no more invisible than an elk drinking at the river—it sends ripples across the water, carves footprints in the mud, leaves rings at the heart of a tree. Nature offers sundials and calendars at every turn. For example, as I write this it is late summer in North America. Bicycling today through the rolling hills, I noticed Queen Anne's lace (wild carrot) dotting the roadsides along with blue chicory, spiky brown teasels, and purple loosestrife. The wild sweet peas that used to tangle with any available tree, bush, or lawn ornament have withered. So, too, have svelte pink dame's rocket and common daylilies. Every day for weeks I seemed to disturb one female ring-necked killdeer, a stilty-legged low-nesting bird notorious for distracting attention from its brood by dragging a wing as it scurries away. Yesterday and today, no killdeer. Its chicks have flown, as they do late in the season.

Clouds of fireflies were June's festival of lights. Now the moonless nights are dark and fiercely raucous. Cicadas sound like a thousand snakes rattling in unison. An occasional raccoon will butcher a baby rabbit, whose high-pitched screams are spine-wrenching to hear, even though I know raccoons have babies to feed, too. The house wrens, which laboriously nested and fed their ever-peeping chicks, have now flown. The stately magnolia outside my bay window has formed thick ropy buds that next spring will open into brandy snifter–shaped flowers.

In early summer, the brown corduroy of plowed fields smelled like freshly turned moss, manure, worms, and cinnamon. Risen corn now means the season is waning. Somehow the cornstalks always surprise me at summer's end. But I do love the golden tassels that catch the breeze like wayward forelocks. All the cornstalks seem to be pointing up; they remind me of Leonardo da Vinci's portrait of John the Baptist with one hand raised, finger pointing toward heaven. By September the fields will become armies of withered scarecrows, and I'll fret as I bicycle past them, fleet over the earth but disturbed by the fleeing season. Despite rampant summer heat, I find myself counting the buds on the dinner-plate hibiscus: thirty more left. It's hard not to cling, to keep deadheading, to hope summer will meander into December, though I know the plants are exhausted and need to rest their roots in cool soil.

We also respond to the more ethereal traces of time, such as sunlight prowling around the house and climbing in the west-facing bay window at around 3 p.m. Best to let the filmy, semipermeable blind down then, lest our local star bleach the cushions.

Driving through mountain passes, one can see solid time laid out in strata, geologic eras piled like so many oriental rugs. Nature's calendar depends on locale, of course. A crust of ice on a lake in the East signals winter, but not so in the Klondike. In the eastern United States, migrating birds draw check marks across the sky in autumn while squirrels bury nuts and grow fluffy bellies to prepare for winter. But in California monarch butterflies will be invading eucalyptus groves.

One misty, cold morning, I stroll along the gorge trail to Taughannock Falls. The waters are high and violent, and when I cross a bridge to the lookout point near the bottom of the towering falls, the air grows thick with spray and aerosols of mud and silt. Above, time reveals itself in ribbons of rock. The falls spill over a cauldron of limestone, the hard core of eons. But where I stand the canyon is mainly shale, a soft graphite-like rock that is constantly crumbling, powdering, breaking down. Two contrasting types of rock frame this gorge, and when you stand at the base of them you can see clearly, as one rarely can in life, a basic principle of time in nature: what resists and what falls away. The shale begins with water and ends with water in a cycle of construction and destruction. The woods, too, are constantly sloughing. You can see it in the peeling bark, the decaying logs, the crumbling stone. In that setting, I recall a friend who recently died, an ebullient young woman whose foot cancer ultimately killed her. Determined to live heartily, she stayed upbeat, positive; she became engaged, planned a future, married. *Denial* isn't the right word for what she seemed to feel toward the end. She knew she was dying, but resisted with all the energy she could muster, while her body fell away.

Time both races and pools in one's memory. I wonder, do the same neurons convey both illusions? An example of a time pool: The Columbus, Ohio, *Dispatch* ran a poignant photograph, taken at the Dayton Airforce Museum, which was honoring the last surviving pilot of the famous mission to bomb the bridge over the River Kwai. In the photograph an old man poses for journalists, holding a picture of himself as a nineteen-year-old, standing rakishly in front of a plane on which he has painted a sea monster's nose and dozens of bombs. Hat cocked, khaki shirt open at the neck, the young man looks suntanned, slender, and luminous with life. His expression says he is indestructible, he will be nineteen forever. After all, he controls whirlwinds of metal, he can blow up bridges and whole cities. What is time compared to *that?* He doesn't know that his real enemy, the one RAF pilot and World War II memoirist Richard Hillary called the last enemy, lives inside his cells, is unreachable, and will certainly defeat him. He remembers his war years as coated in thrill, a delirious time of irresponsibility and freedom. It's easy to understand why many men cherish the red-alert of war as a high point their lives will never again achieve, a time when the world seemed carefree and savage, passionate and disarming, with nonstop moments of hatred, love, and deliverance.

An example of a time race: When I was in junior high school in Allentown, Pennsylvania, every classroom had a sign beneath a large wall clock that read: "Time passes. Will you?" We students understood that the question was meant to motivate, despite its ambiguity. Most of us would pass junior high school, but what we didn't know then—when we were wont to stash a dead groundhog in a locker before the weekend, or insert Popsicle sticks between the strings of the music teacher's piano, or brew 100-proof alcohol in the chemistry lab—was that time would pass faster than we wished, faster than our brains could reckon, as fast as fatal rocketry. Some students would die in the war in Vietnam, and others would fall victim to a mysterious virus or bacterium. But at the time all we thought about was growing up—and not gradually, but in a single breakthrough of height, breasts, and vocal cords. We longed for the fine lines of worldliness visible in the close-ups of glamorous movie stars. We traced our stages of growth with bar mitzvahs, sweet-sixteen parties, proms, and other rites of passage. That still happens, of course, as kids move from a jump rope to a rope of pearls. Meanwhile, time leaves a trail of fossils in limestone. Meteor showers guide us through the seasons, as does the waltz of the constellations. Each month a curvaceous moon fattens and skinnies. Lengthening shadows alert us to day's end as surely as sundials. If we're lucky, a long-haired comet may appear for the first time in 4,000 years. When we peer at exquisite nature photographs, as I often do during this season, time stops for precious moments, allowing us to step between tick and tock, while a camera gasps.

Autumn

The true mystery of the world is the visible, not the invisible.

—Oscar Wilde

30

Dawn frost sits heavily on the grass and turns metal fencing into a string of stars. Was that a goldfinch perching in a 100-year-old oak or just the first turning leaf? A cardinal or a sugar maple closing up its canopy for the winter? Seasoned trackers, we stand still and squint hard, looking for signs. On a distant ridge, a yellow patch, surrounded by green, glows like a room where the light has been turned on. At last the truth dawns on us: autumn is stealing into town, on schedule, with its entourage of chilly nights, macabre holidays, and spectacular, heart-stoppingly beautiful trees. Soon the leaves will start cringing and roll up in clenched fists before they actually fall off. The vast green leaf castles of summer will vanish like a mirage. But first there will be weeks of hypnotic colors so sensuous, shrieking, and confetti-like that people will travel for many miles just to stare at them—a whole season of jeweled leaves.

Why do the colors form? They don't, they undress. Soon after the summer solstice (June 21), when days begin to shorten, a tree reconsiders its leaves. All summer it feeds them and they process sunlight, but as days shorten the tree gradually chokes off its leaves by pulling nutrients down to the trunk and roots, storing them there for winter. Spongy cells form at the leaves' slender petioles, then scar over. With little nourishment, the leaves can't manufacture the green pigment chlorophyll, and photosynthesis stops. Animals can migrate, hibernate, or store food to prepare for winter. A tree has fewer options. It hibernates by dropping its leaves, and at the end of autumn only fragile threads hold leaves to their stems. Turning leaves stay partly green at first, then reveal splotches of yellow and red as the last chlorophyll gradually breaks down. Dark green seems to stay longest in the veins, outlining and defining bright colors like the bevels holding stained glass. All summer long, used-up chlorophyl is replaced. But in autumn, because no new pigment arrives, the green screens dissolve and colors leap. Camouflage gone, reds and oranges seem to arrive from somewhere, but in fact they were always present, a vivid secret hidden beneath the green plasma of summer. Coleus and other plants that start out with variegated red, yellow and even purple leaves are also in costume: they're green at heart, busily producing chlorophyll like their neighbors, but their bubbling colors mask the green.

European maples don't achieve the same flaming reds as their American relatives, which thrive on cold nights and sunny days. Warm, humid European weather tends to turn leaves brown or mildly yellow. Thanks in part to a robust climate, the most spectacular fall foliage occurs in the northeastern United States, in eastern China, and in Canada. Anthocyanin, the pigment that gives apples their red, and also turns leaves red or red-violet, is produced by sugars that remain in the leaf after the supply of nutrients dwindles. Unlike the carotenes, which color carrots, squash, and corn, give roses their fragrance, and turn leaves orange and yellow, anthocyanin varies from year to year, depending on the temperature and amount of sunlight. The brightest colors appear in years when the fall sunlight is fierce and the nights are cool and dry (a state of harmony travel guides find vexing to forecast). This is also why leaves appear sharp and clear on a sunny fall day: the anthocyanin flashes like steel.

Elms, ginkgos, hickories, aspens, bottlebrush buckeyes, cottonwoods, weeping willows, and poplars all turn a radiant yellow. Basswood grows bronze, birches bright gold. Water-loving maples put on a fireworks display of scarlets. Sumacs redden, as do flowering dogwoods, black gums, and sweet gums. Though some oaks yellow, most turn a pinkish brown. The meadows and farms change color, too, as cornstalk tepees and bales of hay dry in the fields. One slope of a hill may be green and the other already in color, because the southern hillside gets more sun and heat than the northern one.

As with so many of the sensations we adore, leaf colors don't have any special purpose. We've evolved a response to beauty, and fall leaves sizzle with the flames of sunset, sparkle of spring flowers, or shuddering pink of a blush. Leaves and flowers and animals alike change color to adapt to their environment. But there is no human reason for leaves to color so beautifully in the fall any more than there is for the sky or ocean to be blue. It's just one of the haphazard marvels the planet dishes out every year. We find the shimmery colors thrilling, and in a sense they dupe us. Though colored like living things, they signal death and disintegration. In time, they will become fragile and return to dust. They sublime from one beautiful state to another, much as some religions tell us we will. As leaves lose their green life but bloom with urgent colors, the woods grow mummified. Nature becomes carnal, mute, and radiant.

We've always called the season "fall," from the Old English *faellan,* to fall down, which leads back through time to the Indo-European *phol,* to fall. The word hasn't really changed since the first of our kind needed a name for this seasonal metamorphosis. Then there is that other fall, the one in the Garden of Eden. Adam and Eve concealed their nakedness with a fig leaf, remember? Leaves have always hidden our awkward secrets. Fall is the time when leaves fall from the trees, just as spring is when flowers spring up, summer is when we simmer, and winter is when we whine from the cold.

For children, flurrying leaves are just one of the odder figments of nature, like hailstones, or snowflakes. They love to plunge into soft, unruly mattresses of leaves, tunnel through leaf mounds, and hurl leaves into the air. Walking down a lane overhung with trees in the paint-splatter of autumn, one forgets about time and death, lost in the sheer delicious spill of color.

A light breeze and the leaves are airborne. They glide and swoop, rocking in invisible cradles. Fluttering from yard to yard on small whirlwinds or updrafts, swiveling as they go, they are all wing. But how do they let go? As a leaf ages, the growth hormone fades, and cells at the base of the petiole divide. Two or three rows of small cells, lying at right angles to the axis of the petiole, react with water, then come apart, leaving the petioles hanging on by only a few strands of xylem. Then even those disintegrate, and the gliding begins.

Tethered to Earth, we love to see things rise up and fly. Soap bubbles, balloons, birds, fall leaves. We especially like the way they swoop, rock, and swing as they fall. Everyone knows the motion. Pilots sometimes do a maneuver called "a falling leaf," in which the plane loses altitude quickly and on purpose by slipping first to the right, then to the left. The machine weighs a ton or more, but in a pilot's mind it is a weightless thing, a falling leaf in, say, the Vermont woods where children play. Below her the trees radiate gold, copper, and red. Leaves are falling, although she can't see them fall, as she falls, swooping down for a closer view.

In time, the leaves leave. But first they turn color and thrill us for weeks on end. They crunch and crackle underfoot. They *shush,* as children drag their small feet through heaps along the curb. Dark, slimy mats of leaves cling to our heels after a rain. A damp stucco-like mortar of semidecayed leaves protects the tender shoots with a roof until spring and makes a rich humus. In leafy mounds, an occasional bulge or ripple signals a shrew or a field mouse tunneling out of sight. In the front yard lies a fossil stone with the imprint of a leaf, long since disintegrated, whose outlines remind us how detailed, vibrant, and alive are the things of this earth which perish.

31

The adult hummingbirds, Ruby and Greensleeves, left on September 5, as if they had a train to catch. Not together; hummingbirds are loners. For weeks they've been guzzling food, fattening up, and preparing for the arduous flight over sea and city, driven by a restlessness they could feel in their blood, a hormonal surge reliable and mysterious, that persuades them to hit the airways and fly until they reach their winter home on the Yucatán peninsula. They will know it when they see it. Fortunately, there won't be Aztec warriors waiting. The Aztec war god was a hummingbird, Huitzilopochtli, which meant "shining one with weapon like cactus thorn." The male birds, whose feathers glittered like jeweled armor, were ferocious and quick, as the warriors wished to be. Aztec rulers wore ceremonial robes made entirely of iridescent hummingbird feathers, which caught the sun and bathed them in supernatural lights. I can't imagine how many hummingbirds they must have killed to make one robe.

When I picture Ruby's small wings beating nonstop for days and then again for 500 miles over open ocean, his small heart sprinting, I wonder how he'll make it. Where is Ruby now? Will he stop to rest along the way, in Charleston perhaps? Will he be able to avoid head winds and hitch a ride on balmy currents? Will he return next year to this same feeder, or will he die somewhere in the Central American rain forest? And what of Greensleeves and her baby? Will they have flown to the same sunny spot or parted ways as Gizmo begins his own life story? Will Gizmo visit my yard next summer?

Although I continue putting up fresh hummingbird nectar, I feel a mixture of loss and relief. Every morning and evening they arrived like magical incantations. I'll miss them. But, as with the wrens, I'm also relieved they survived. I don't know what hummingbirds feel, or how they remember. Will they recall this specific garden and return to it? Probably so. Do they remember the succulence of certain flowers, the hidden comfort of certain nests? Does the mother remember she had a chick? Or is their memory sketchier, largely navigational? Hummingbirds may live a dozen years, and I can just picture Gizmo's return next spring in the full plumage of an adult male.

They didn't need to leave yet, there's still plenty of nectar and the days are hot and steamy. Snapdragons still bloom like a spill of colorful marbles, the bright orange trumpet flowers continue climbing the fence, black-eyed Susans create sunbursts in every direction, the red geranium trees have thirty new flowers, and the roses are galloping through a second bloom. Othello now has eight canes twelve feet tall, and seven more that stop at five feet—all pounding out blossoms. Most roses make a modest bloom at summer's end, but some get lathered up for a full pageant every bit as spirited as their spring show. Mister Lincoln is producing flowers big as cereal bowls, and Colette has doubled in size and begun embroidering a fence with dozens of pink flowers.

But fothergilla leaves have turned the color of blush wine. Summer's tall green-and-purple meadow rue has formed beads of delicate seedpods and saffron-yellow leaves. The burning bushes are aflame. A few magnolia leaves have dried to the tan of old parchment. On the large sycamore in the front yard, leaves are pausing at a pale lime-green stage, each day yellowing a touch more. Few sights are as beautiful as ornamental grasses, their many shapes and colors clumped along a fifty-foot border, waving flags like a parade crowd. "Consider the grasses," A. S. Byatt writes in *Still Life* (1985), "so carefully distinguished from one another. They are little figures of speech. *Glastridium*—nit-grass, from *gastridion*, a little swelling. *Aira*—hair-grass, from *airo,* to destroy (a destructive grass). *Panicum*—or panick-grass, from *panis,* bread, because the seeds of this grass can be milled and eaten." They also create a colorful mosaic where the tame meets the wild in one long line of plumes, green cushions, orange-brown spires, yellow zebra stripes, tawny upside-down brooms, golden sheaths, blue-green spikes, clouds of apricot—all swaying in the lightest breeze, bobbing and bowing and lasting for six months. Something softening about the grasses, a remembrance of prairie, blurs the edge of the property and offsets the blank black macadam of the driveway.

Among the grasses, rudbeckia and other yellow sunflowers tower up through tawny leaves. Armies of lanky pink Stonecrop sedum stand sentinel in the garden. For a few weeks, the world was thick with the floral spires of blue lobelias, all of which have dwindled to raggedy poles. This is a time of paradox, surfeit and famine, when summer and fall perch on the same vine. When the leaves start leaving, can the birds be far behind?

I have a seasonal calendar to follow, too, but I must confess I'm not a master gardener by a long shot, nor even a particularly expert one. I've had some garden experience, read heavily, and often been on the losing side of trial and error, and I'm happy to share whatever I've learned. But I lack a certain stoicism that professional gardeners cultivate. For example, as I said, the roses are in a second wave of bloom. On my rounds today, I brought in one stem holding five sweet-smelling sunrise-colored Abraham Darbys, a giant red Cardinal's Song with frilly yellow tassels at its heart, a tiny flame-orange Charisma on an unusually long stem, a close-fisted pink Paris de Yves St. Laurent, four perfectly formed tangerine-colored Lasting Peace, one red-and-white-striped Scentimental (whose loosening petals look like lingerie slipping off a bureau), a canary-yellow Climbing Golden Showers, and two talcy pink Summer's Kisses. I haven't arranged them yet. I left them in a jug on the kitchen counter and soon will spend an admittedly indulgent amount of time choosing the right vases to contain their charms, filling those vases with tepid water, adding a few drops of lemon juice to help fight bacteria, stirring gently like a martini, clipping off any leaves that would dangle underwater and rot, before I finally trim the rose stems at an angle and place them in their tiny lagoons. Then I will inhale them, one after another, coming up for air in between, enjoying their subtly different scents. Sometimes I just stand and feast my eyes on them tirelessly, as if they were children home from boarding school. I don't apologize for being a poet who loves roses. But I lack the stoicism to do the right thing and stop deadheading the roses now. I should let them form hips and gradually prepare for winter. I won't fertilize them again this season, but I can't resist bringing in beautiful homegrown roses. Cutting flowers is really a form of pruning, and by doing that now, late in the season, I'm confusing the plants into thinking it's still summer and they need to bloom. A rose busily minting flowers and leaves late into the fall will get a rude shock when the first frost hits. What I *should* do is feed the roses a nitrogen-poor 0-10-10 fertilizer, perhaps, to encourage root growth, directing the plants' energy away from flowers. I should allow the roses to fade gently into hibernation, forming hips and seeds, before they go dormant. Soon they will move food into the canes and roots, storing it there for winter. I must let the leaves wither and fall. If I strip them off too early, the plants might not get enough nourishment to stockpile for the long winter ahead. If I overprune the canes (so that they won't get torn by ice and winds), the roses may think it's a serious spring pruning and mistakenly rev up for the new season's growth. I know what I should do. But not this week, nor the next probably. After all, they're still colorful as party balloons. Every morning I find several new surprises reclining in the beds, climbing up the fences, clinging to an obelisk, leaning on their canes. It's like greeting a troupe of acrobats. Of course, I never want their bloom to fade. Maybe I'll let them swill sunshine a little longer.

Several bees keep circling the hummingbird feeder, landing, trying to get the sugar water, which they can smell but not reach. Suddenly a hummingbird arrives. It's little Gizmo, looking plumper and in a bad temper. Swirling and diving, he chases them in a real Battle of Britain display. We used to say of ranch quarter horses that they could turn on a dime and give you a nickel change. Gizmo swerves and spins and soars all in one move, hazing the bees away as if they were ornery cattle. His rapier bill hasn't grown to full size yet, but that doesn't matter. He's ferocious. After a few rounds, the bees buzz off and he drinks his fill. You understand, he doesn't *need* to chase them away. They're after the sugar, not him, and there's plenty of food to go around. But male hummingbirds are incredibly territorial, and the bees just piss him off. Personally, I'm happy to let the bees enjoy this last sweetness at the end of their lives, whereas Gizmo has many more summers to savor. Why is he still here, when both parents have left? Since this will be his first migration, he might not recognize the siren call to travel yet, or perhaps he's lingering to gather strength and fill his fat stores before setting off on what will be the worst journey of his young life.

32

Everyone is sneezing. The back of my nose tickles, too, and I feel like I'm swallowing a ball of rubberbands. I hope it's just ragweed pollen, but fearing I might be coming down with a cold, I've been sleeping in the garden room the past two nights, in a sort of bell jar. It's wonderful waking up surrounded by the garden, at ground level, as my eyelids open like curtains and dawn starts seeping through the woods. This morning, just before sunrise, a loud cardinal serenaded with a pure, startling song: *Write it down. Write it down. Redo. Redo. Redo!* it sang over and over. Perching high in the quaking aspens, it hid from sight, magnificently musical and insistent. How long had it been up?

I've been drinking hot echinacea (aka coneflower) tea , a traditional boost for the immune system. We've grown so far away from our food and medicine sources that it feels a little odd to be sipping tea from what grows decoratively in my front yard, where a stand of pink coneflowers offers solace to passersby. A sneezing fit.

Pollen, not a virus, is the culprit. Autumn can be a pollen nightmare for the allergic, though of course it's vital to plants busily breeding before winter's stem-breaking gales and lockjaw of frozen earth.

Once you see a pollen grain under a microscope, you forgive the mischief it makes in your sinuses. Beautiful little satellites, spiky, and born to cling, pollen scratches by design. Few are round, some are oval, others sport sails. When fields of feathery goldenrod bloom in autumn, it's usually blamed for allergies. But brilliant color is its alibi—colorful flowers lure and are pollinated by insects. Goldenrod waits to be visited. Along with asters, it provides honeybees with a last source of nectar in autumn. No, it's the shy plants one needs to keep an eye on, drab ragweed, for instance, which usually grows among goldenrod and produces small, light pollen that sails well on the wind. Relying on the buckshot technique, ragweed launches 1.6 billion pollen grains per hour. That sounds like a lot, but many grains will be lost in transit, and pollen is cheap. Wind-pollinated plants don't have to design alluring colors or create nectar as bait for insects. They just flood the neighborhood with seed. Humans produce millions of sperm, too, which is why one teenage boy could populate an entire nation, and many have tried.

Some low-lying plants prefer wind pollination, but it's really ideal for trees, whose seeds can glide around in clever little whirligigs or catkins. Spreading seeds is only half the job; plants also have to capture pollen, and that can be especially challenging. But some plants have devised ways to harvest the wind as cleverly, I think, as our inventing windmills, windsocks, and airfoils. Consider the pinecone. From which direction will the pollen-laden wind be blowing? It doesn't matter. The pine tree evolved cones as tiny turbines with blades that catch pollen from any direction and distribute it in a wind cloud around all the blades. Where swirling and still air meet, currents form that guide the pollen to its target. Or consider the jojoba plant. Above each of its hanging flowers, two leaves grow at the precise angle to channel the wind down to where the female ovules wait for the pollen. Surrounded by an ocean of air, flowers have learned to exploit the wind in novel ways. So I guess I shouldn't be surprised, in autumn, to find the air thick with seeds we breathe in and snort out, without quite knowing which growth process we've joined.

I can hear the neighbor children laughing like faucets as they plunge into leaf piles. Oddly shaped squashes decorate many doorsteps. At Ludgate's, a favorite roadside produce store, I buy fragrant basketfuls of Reliance, the seedless red grape prized for its taut skin and sweet chewy flesh. The sky is a Mediterranean blue, and the temperature benign. In shade, one shivers; in direct sun, one sweats.

Autumn also means garage sales meandering through the neighborhood and drawing a steady stream of treasure hunters all day. I'm amazed by what people will cart off. It's like scavengers on the Serengeti. When they see a carcass, however old and frayed and picked-over, they feel the urge to drag it home.

One day, when a house across the street was holding a yard sale, I put a rusty plant box and a tower of ratty plastic hanging planters at the foot of my driveway, with a small card that said "Free." By the time I walked back to the top of the driveway and turned around, they were already gone, silently loaded into an SUV plastered with bumper stickers, only three of which I could read as the car pulled away:

WARNING: THIS DRIVER JUST DOESN'T GIVE A SHIT.

IF IT HAS TIRES OR TESTICLES, IT'S TROUBLE.

I'M SO GOOD AT MOLLIFICATION I'VE BUILT A GROVEL PIT.

I thought the last one particularly good, and was sorry I couldn't peruse the rest, disappearing slowly around a curve, hauling a load of goodies that are some folks' trash and other folks' windfall.

A butterfly sitting on a coneflower is turning slowly in place, sipping nectar as it pirouettes. Its wings open and close in slow, regular beats as it turns. Whenever the wings open they look like a

Cupid's-bow mouth. Closed, seen from the side, they are the stylized angel's wings (minus head and body) so popular now as decoration, by relying on *synecdoche* (the part stands for the whole); just show us the wings, and we know it's an angel. Why this butterfly has to beat its wings at all while it perches I don't know, unless a to-me-intangible breeze is enough to tumble it from the nectar pools. Although I can't see it from where I'm standing, I know the butterfly has a long coiled tongue that unrolls like a party favor when it wants to sip. And it has tiny grappling hooks for feet. So you'd think it could stand still on a flower. But maybe it's enjoying the hot sun on this coolish day and using its wings as solar panels to gather heat. (For it to use its flight muscles, the air temperature must be at least 55 degrees.) Not too much heat, just enough to feel comfortable, the amount gathered by slowly opening and closing its wings. I've often seen animals do this simple-looking but intricate thing, adjusting their thermostat from moment to moment. We do it, too, of course, which is why sweater "twin sets" are so popular and a man might wear a short-sleeved shirt but carry a jacket. The alligator version of this is to bask with its tail and one leg in the water, and the rest of its body on the sunny bank. Alligators prefer ponds covered in what looks like green scum but is actually a warm blanket of the world's smallest flowers.

It's a heavy, overcast day that's palpably humid. The air feels like cloud. Everyone is praying for rain so they won't have to water their parched gardens. Some of the coneflowers have begun drooping, which tells me what my chore is for today. The lively being of the garden is crying out to be fed.

33

Deadheading asters will teach you humility. It can be a peaceful pursuit as you scan and snip, scan and snip. The new buds form so close to the flowers that you can't deadhead several at once. You need sharp pointed scissors and the patience of Penelope. While deadheading in the long row of asters along the driveway, I notice something out of the corner of my eye and turn to see a large adult raccoon sitting on the driveway by my wheelbarrow, inspecting a bag of plant food lying on the ground. When I walk toward the raccoon, it staggers a little, and then a lot, and I stop once I realize how sick it must be. Turning, it careens drunkenly along the side of the house to the bottom of the tall oak in which it nests, but it's too weak to climb. Exhausted from that small effort, it lies down for a few moments, then drags itself slowly into the woods. I phone my neighbors to alert them and their children, then the SPCA, who will be dropping off traps in the hope of capturing the whole raccoon family and testing them for rabies. I'm most concerned about the children in the neighborhood, but dogs and cats can contract rabies, too, and pass it along to their owners. A horrible illness, rabies mysteriously killed Edgar Allan Poe in 1849, after first making him delirious for three days (which is the normal course of the illness). Staggering around Baltimore, he finally collapsed on the street and was rushed to a hospital, where he improved and then relapsed, dying from a bite he may have received as much as a year beforehand, perhaps from one of his cats, which had contracted it from a rabid raccoon.

Although it's still brutally hot in Ithaca, temperatures in the 50s are forecast for later this week, and the maples are a heart-stopping red. I'm all set to dash out on chilly nights with my caravanserai of tent poles and pink sheets, desperate to extend the growing season, but I know it's only a matter of weeks. Harsh. I hate closing down the garden, but the flowers need to go into suspension and rest, as growing things must.

The chocolate-leafed snakeroot plants I put in this year have clusters of tiny white flowers, and the contrast of the reddish-brown stems and leaves set off by snowy flowers is thrilling. In one bed, Joe-Pye weed towers with mauve blooms. Scarlet coleuses have fattened into brilliant bushes. Elephant-eared caladiums show blood-red veins against mint-green leaves. The peppermint-striped rosebud impatiens have grown three feet tall. Purple astilbe feathers are just finishing: the garden is putting away its feather pens. White daisy-flowered asters (the roadside variety) loom over compact deep purple asters. Miniature yellow grasses form tuffets, pink windflowers flutter in the breeze, and drifts of black-eyed Susans add shocks of yellow. A large brown pot, tucked into the foliage, contains a giant yellow mum, heavily in flower. Two yuccas wave yellow-and-green-striped leaves that look like garter snakes. Showy sedum are blushing pink knobs. And a dozen asters of all sizes and colors surround a lavender rose of Sharon at the heart of the bed.

Sitting quietly, savoring the panorama of bloom, I begin to smile. I don't like summer to end, but this fall show is worth waiting for and spending time with. It has its own themes and uniquely beautiful plants. The asters are a revelation—they assume so many shapes and sizes. Some are barely visible on long trailing stems, others look like bold pink daisies, still others top stiff tentacles, while some are low compact bouquets of fluorescent fuchsia. Surely I need more asters. A favorite fall flower, asters are named after Astraea, the goddess of innocence, who, according to Greek myth, encountered so much sin among mortals that she metamorphosed into the constellation Virgo. Zeus was fed up with human sin, and in a punitive mood he unleashed great floods upon the earth, killing all but one devoted couple, Deucalion and Pyrrha, who, in the cataclysm and its aftermath, became lost. Taking pity on them, Astraea cried stardust tears, which fell to earth and became asters; then she used a trail of starlight to guide them to safety.

Despite their sad myth, I think of asters as explosive bursts of bloom at summer's end, a plant with dozens of faces, each more spirited than the next. We've got a solid month yet before hard frost, and even then the ground won't freeze for quite a while, not until December. Anyway, soil freezes from the top down, so planting can continue a little longer. Supposedly, asters are a remedy for being bitten by a mad dog, but you have to make a poultice of the flowers and "old hog grease," and I don't know anyone with old hogs—or mad dogs, for that matter. Most of my lavender asters have returned, and I've added new ones. Holding the face of one between my first and second fingers, I remember a poem written by Gottfried Benn, a German doctor who worked in a morgue during the first half of the twentieth century. All of his *Morgue* poems are intriguing, but it's "Little Aster" (1912) which comes to mind. The translation that follows, without rhyme, is by Babette Deutsch, but Benn's original has many full and half-rhymes.

Little Aster

A drowned truckdriver was propped on the slab.
Someone had stuck a lavender aster
between his teeth.
As I cut out the tongue and the palate,
through the chest
under the skin
with a long knife,
I must have touched the flower, for it slid
into the brain lying next.
I packed it into the cavity of the chest
among the excelsior
as it was sewn up.
Drink yourself full in your vase!
Rest softly,
little aster.

In that curious little poem, two passions vie: the cold-blooded, matter-of-factness of the surgeon who performs autopsies deftly and through habit, and the romantic man who sews the flower into the corpse's heart with the same delicacy and assurance he previously used to cut out the tongue and palate. Because it was a theme of his in other poems, I think he meant to emphasize how we moderns (dedicated to science) and the aster (a symbol of nature's gratuity and abandon) rub against each other like two fists. There is barely a pause between the aster's tumbling in with the brains and the doctor's packing the flower into the chest cavity. In one rough gesture, the aster moves from mouth to brains to heart, providing the truckdriver with another limb, turning his heart the violet color of blood, and reinfesting his dead body with myth, by conjuring up all the old ballads in which flowers grow out of lovers' hearts. The aster also finds a kind of salvation in its new vase. Life can sustain itself for a while, in all its beauty, even on death and excelsior (shredded trees). And what of Dr. Benn, who secretly packed it in the chest? He believed the hospitals catered to the wrong diseases, that man's real affliction was being locked out of the healing processes of the seasons. In "Little Aster," he commits feeling as if it were a criminal act, with a starry aster as its symbol. I really haven't much excuse, practical or poetic, for buying asters, I just crave them at this time of year, for the way they dust the garden with life in the thin days of summer, for their effusiveness, frost-resistance, and fruity colors.

Driving to a local nursery, I idly plan how many tall, rangy asters I can cram into my Toyota hatchback. Quite a few really, more than in any other car I've owned. I'm sorry we didn't learn garden math and geometry in school. "If the back of your car is 4 feet wide, and 3 feet from bumper to back of the front seats, and an average of 3 feet high, how many 8-inch pots of asters 4 feet tall by 2 feet wide will fit if (1) you lay them side by side, (2) crisscross them on top of one another, or (3) arrange them head to foot?" When, I wonder, did I stop thinking of cars as transport and begin thinking of them mainly as garden wheelbarrows?

Suddenly I feel affectionate about bugs. Not aphids or Japanese beetles or slugs or mites or boring caterpillers. I remember my first car, an ancient, clapped-out, secondhand VW Beetle, nicknamed a "bug," I suppose because of its rounded carapace. By the time I bought it—for $100 from a student—its mint-green paint job had a sickly white pallor. It was the shade of green left on one's skin by old copper. At nineteen years old, I had been driving for five years, but not ambitiously, and I'd never handled a stick. Too terrified to shift, I stayed in second gear for a month. Every time I tried to shift up into third—which meant muscling the knob rough and fast through an exaggerated H—I heard digestive sounds I didn't think cars were supposed to make and I quickly lost my nerve. Yes, the engine did roar when I went over 40 mph in second gear, but noises from the rusty muffler added to the general ruckus, and because the springs were all shot, for the most part it sounded like I was driving a busted dinette set.

The inside of the car was tattered, as if someone had been trying to claw his way out. Perhaps it was psychosensitive and had been driven to hysteria by its owner's neuroses? I couldn't be sure. But I did love the freedom that car gave me, carrying me out to the countryside to walk among the wildflowers and beside farm fields of wheat and corn. I can still remember the mixing smells of honeysuckle, wild onions, and freshly mown grass, as I would drive slowly down dirt roads, admiring the farm gardens and trees old enough to shade most of a house. I wish I had snapshots of each acre—the clumpy rows of wet, ripe wheat, each head resting on its own arms in odd brilliances of light; the braided Alleghenies so lush I half expected to see them dripping with liana vines and vanilla beans and parrots chattering in the catch-fire heat of the Madre de Dios; meadows of Queen Anne's lace blowing like thousands of of doilies; horizonless rows of corn; and especially, now and then, a stand of ancient slope-shouldered willow trees, whose huge, drooping limbs always evoked baboons.

In time I found the car's percolating noises as comforting as a cup of tea. I would have given it up for love, I suppose, but not money. Even so, it wasn't safe at high speeds. Once, when I took a turn a little too fast, the rear end swung out, the car spun around and bounced off a rock wall, coming to rest in the center of the highway. A pause as I realized I was alive. Then all of the fenders and doors dropped off in one loud clatter. Though alarming, that incident wouldn't have persuaded me to give up my beloved VW. The windows could be taped over; the fenders and doors could be strapped back on with rope. No, it was only when a friend borrowed the car and ran it without oil until the engine seized that I reluctantly parted with it. So it went the way of all metal, to that great memory junkyard in the sky.

It was doomed because there's no way it could have served my passion for gardening. A few years ago, I thought a car with a hatchback would be enough for plants and skis and humans, and it was. But now I find myself shamelessly applying the critieria of gardening to everything. Forget comfort. Forget style. Forget economy. Will it serve the needs of the garden? That is the question that starts to narrow a gardener's life. In the vortex of one's obsession, the landscape starts to spin, and objects one counted on as solid, stubborn givens pull free and swirl into the center of the addiction. Gardening an addiction? I suppose life is the addiction, gardening one of its better drugs. "Little by little, even with other cares," Henry James confesses, "the slowly but surely working poison of the garden-mania begins to stir in my long-sluggish veins." Mea culpa. I'm fond of a gray T-shirt I picked up at a fisherman's store in Skaneateles. It shows a fisherman tying colorful flies, above which it says: "So many lures . . . so little time."

Today I will also cram my car with chrysanthemums reputed to be "hardy," which means they've "wintered over"—survived at least one winter outside. The chrysanthemum lottery happens every autumn, when trusting souls purchase bright cushiony mums, believing they will return next year, not bite the frost (if they survive the slugs). Sellers a little casual with the truth might *say* the mums are hardy when they're spanking new and from a greenhouse.

It's hard enough to endure plunging temperatures if you can find shelter. Plants have evolved ingenious ways to survive winter. Many strategies will work, and here are just a few: Some plants hide out underground—as roots, bulbs, and tubers crammed with food—until it's safe to grow leaves again; many plants secrete alcohols and sugars as a kind of antifreeze to lower the temperature at which cell walls would burst; lichens dehydrate over the winter; some plants grow low to the ground to avoid windchill; others create their own microclimates underground; some flowering plants (like mountain laurel) grow hairs along their stems and fruit as insulation; arctic flowers often use large petals as sun traps; some just sink their roots deeper; plants that live in extreme cold (such as red algae, which can grow on top of ice) sometimes use a color like red to convert light into heat. The plants in my yard are luckier—I cover some with pink sheets for a short spell, then bundle them in parkas of canvas and pine needles and dry leaves. It's a symbiosis as dear as the one between dogs and humans. I nourish them and they nourish me.

34

Roses are such wild creatures that I keep them in cages and pack them solidly with fall leaves. It's a beautiful effect but also practical. When they're caged their leaves don't tangle, and they look orderly, contained, predictable, plumper than usual, sentry-like. I cage other things, too.

Each autumn, after planting new tulip bulbs, I cover them with chicken wire and spread mulch on top. Another good way to protect bulbs from squirrels is to cut up aluminum pie plates, fold a wedge over on each, and plant them as sun-strobes, to jolt squirrels away. Unfortunately, I don't like the migraine-makers either, so I sink cages. This morning, I watched an irritated red squirrel frantically trying to pull a tulip bulb up through the chicken wire, only to be stymied with each attempt. When I walked outside, she ran off. I covered the bulb again with dirt and mulch, but as soon as I went indoors she returned and started digging. This continued for some while. Around midday, she returned once more, sat and puzzled over the spot for quite a while, dug at the mulch, reached through the chicken wire and began rotating the bulb like a soccer ball. *Determined little thing, aren't you?* I thought, feeling pretty smug about the tulip bulbs, some of which I'd actually buried inside chicken wire treasure chests. If this were a cartoon, smoke would have been rising from her brain.

Finally, she switched her attention to a yellow gourd I'd left out in the hope it might distract her. Before long she had gnawed into it, eaten some of the flesh, and begun removing and shelling the tasty seeds. At least she was leaving my tulips alone, but I knew she wouldn't give up easily. The unobtainable in a cage—it's irresistible. She could see it, touch it, even revolve it, but not grasp it firmly, not savor it, not profit from its richness. On the other hand, she showed remarkable focus and determination, two traits that the earliest humans must also have had. It made them successful hunters and gatherers for thousands of years, in that limbo before *Homo sapiens* appeared with creative minds and lots of gossip. From our perspective, the red squirrel is "wildlife," and human beings, including all of our previous versions, no matter how primitive, are not. For the time being, anyway.

Will wildlife still exist in future days? Or will we have driven most animals to extinction, burned the rain forests, and invaded every niche, leaving death and destruction in our wake? The doomsayers warn us that plants and animals are vanishing faster than we can prevent, in many cases before we've even had chance to name them. "I'll burn that bridge when I come to it," seems to be our motto. Of course, the fossil record teaches us that extinctions are normal. Most of the life-forms that once lived on the planet have gone extinct, including some of our own close relatives. It's the current pace of extinction that's so worrying, and our role in triggering it. We've become much better at transforming nature than at understanding it. Other animals can't keep up with us. And just to make sure they don't, we banish them from our daily lives. True, we keep pets, but we treat them like specially exempt children who aren't expected to grow up.

For most of us, animals live far away, in a park, zoo, forest, or ocean we visit from time to time. If we continue to exile nature to the edges of our cities and our thought, we will dramatically change our idea of the wild. Over the past millennium, that has already happened several times. Look at the vocabulary of our earliest ancestors (visible in Indo-European); there we find *kailo-*, which has come down to us as "holy," a word that meant the healthy interrelatedness of all living things. Not only were the gods in their heaven and all right with the world, but humans, plants, and animals were woven together in the seamless fabric of nature. Touch even the fringe of that idea and distant parts reverberate.

According to the Bible, the earth was given to humans as a lush estate we might regard as a pastime or brimming larder—to enslave, plunder, or reform—according to our need. From time to time, civilized folk have pictured the wilderness as a nightmare region of chaos and horror where fanged beasts crouch, ready to strip away one's breath. Most people still imagine the wilderness as desperately Other, a nonhuman realm where it's best to travel with weapons and be guided by an us-against-them mentality. But that idea is far more dangerous than any cougar. It strikes at the heart of all our manners and dreams and bankrupts our sense of identity.

What will become of the wild that lives in us, our own private wilderness? We are wonderfully animal in our habits, fears, and family ways. True, we are different from other animals, just as a stingray is different from a wildebeest, and that difference is probably profound, because our senses and mental fantasia have evolved to cope with unique threats and enticements. But our instincts? Our motives? Our biology? Our basic needs? Pure animal. Watch a bird or squirrel long enough and you'll see all-too-familiar struggles over status, resources, and attachments. Pretending we're a ruling elite isn't good for the planet, but it also isn't good for our own sense of origin, belonging, wholeness, and spirituality. Nature is where you go for replenishment, to reconnect with life's basic forces and fevers. One feels that somehow it doesn't include socks, cars, or colleges. But that's not true. Nature, by definition, contains all living things and all of their dramas—mental or physical. Technically, then, nature embraces TV shows, mascara, and baseball games, just as it embraces talons and turf fights. The dinner date is just another form of courtship feeding, even if it means wearing pantyhose or cufflinks to a glass-and-steel eatery. Sometimes people picture technology as unnatural, and worry that we might lose our humanity in a future of zealously designed machines, but this also implies an adversarial attitude about nature. Nature isn't an alien force; nor does it surround us. We *are* nature, and our cities and inventions, like termite mounds, are part of its complexity.

Impatient by instinct, we've used technology to sidestep evolution: instead of waiting for our eyes to evolve, we've augmented them with telescopes, microscopes, and MRIs. The technology we invent to extend our senses may feel other and nonliving, but it's as much a product of our animal nature as a burrow is of a wombat's. We just don't like to think of it in that way, because doing so seems to trivialize us. That's a fundamental problem we have with nature. We believe it is beneath us, or rather behind us. We believe we are evolution's goal, and that which is animal is dirty, low-class, immoral. We think we can somehow shed our animal nature. We feel valuable only if we can master something, or someone, and unfortunately we like to test that status by regularly proving our dominion. Thus we dam rivers, we level forests, we hunt animals to extinction, we move mountains.

In Zen-like moods, I try to curb my own strivings, but everything in the garden—on land, water, and air—is busily competing for essentials. The traditional macho threat of "This town isn't big enough for both of us!" isn't an idle one. It's just the way humans express a simple biological truth that all plants and animals share: two species can't occupy the same habitat forever (this is known as "the competitive-exclusion principle"). But one of the gunslingers might find a spot unthreatening to the other. Every habitat includes many small niches some plant or animal is busy homesteading. Consider specialists in college departments. Two professors in the same field might indeed find that the department isn't big enough for their equally strong egos, and then one might take a job elsewhere. But it's more likely they'll develop niches, emphasize different areas of expertise. One might accept living in the shade of the other. Both can keep a foothold if they don't occupy exactly the same niche. If there is only one of a plant or a person in a habitat, it will loom larger than if it were surrounded by many of its kind. Richer soil is usually taken over by a dominant species, whereas there's more variety in poorer soil. Also true in college departments. These similarities fit so easily because they concern ecosystems, the coexistence of life-forms. The same forces that rule plants at times rule human behavior, and for some of the same reasons. We're so caught up by the forces, we're not aware of their features, or that we sometimes act much like newts or rhododendrons. Nothing wrong with that. I happen to like newts and rhododenrons. But it could allow us to feel a little more amusement about ourselves and the elaborate ways we disguise our basic strivings.

How we fit into the natural world, what sort of beings we wish to be, and the role we might play in shaping the future of the planet are all questions we'll be forced to consider one day. We will need to consider them because Earth's resources are rapidly dwindling at precisely the same time that our power over other life-forms is growing dangerous. We may be compassionate beasts, but we're also bullying and destructive. That doesn't mean we should have a *laissez-faire* attitude about our yen for violence. Intriguing research in evolutionary psychology will help us understand a little better the way our animal nature informs our everyday behavior. Perhaps people will come to understand the direct relationship between their actions and the environment. It's no use protesting the trapping of foxes, for example, if you build a house in terrain where foxes once lived and destroy their natural habitat. Although it would have shocked us in the sixties, environmentalists and corporations have begun to collaborate on goals. Thus Merck Chemical protects large stretches of rain forest, where "chemical prospecting" is bound to turn up new drugs. When businesses and conservationists join forces, both prosper.

Perhaps cities will be redesigned to include more living elements and a greater sense of the seasons. In large cities, for example, the constellations are on the ground; those in the sky are rarely visible. Perhaps playing in nature won't be regarded as idle, gardens will be deductible as a medical expense, and health plans will cover our need for green time. I hope many people will have the opportunity to view the planet from space, and return with a fuller sense of what home means. Perhaps our schools will help us marvel at our humble origins in the minute stuff of the Big Bang and the fascinating life-forms we have nonetheless become.

Accepting our own wilderness, rejoicing in it, will be an important step toward learning how best to promote what we love about human nature and curb what shames us. If we can achieve that, a refined sense of ourselves as fundamentally animal, then the world we share—what we call nature—will feel homier, and we'll want to protect it like a second skin.

35

Nature tells many stories. What could be more enthralling than the lurid, though luminous, tale of *femme fatale* fireflies? Or more poignant than hummingbirds dying in their sleep because they live at a flutter their rapid hearts cannot manage? The saga of ancient hominids applying raw talents, obsessions, and guts to triumph in a ruinous world remains an epic of startling ingenuity, sorrow, and humor that spans millennia and includes millions of quirky characters, some better known than others, but each improvising a unique life story.

I often sit in my garden reading books set in other gardens, or tales unfolding somewhere in the grander garden of nature, while hearing birds call and insects sing. The wind rustles hickory leaves as I turn the pages. It's like having a sound track while I read. But I'm troubled by this: In our novels, myths and legends, nature tends to loom as a monstrous character, an adversary dishing out retribution for moral slippage, or as a nightmare region of chaos and horror where fanged beasts crouch, ready to strip us from our breath. In *Ethan Frome, Lord Jim,* or *Moby-Dick,* for example, the voice of the wilderness commands like the Old Testament God. Symbolic jungles or lightning storms chase one's silvery spirit and drive one mad. Massive male animals combat one another as noble but fearsome incarnations of the wild. Even medieval knights face fire-breathing lizards.

What do we fear exactly? I mean after the possibility of a ragged death. I guess we fear becoming as fragmented and plural as the world around us. Plato and many others have written obsessively about the One and the Many, a ratio that seems to speak to us instinctively. The individual amid the herd. The many secret selves contained in a single seemingly coherent personality. The lone outcast against the backdrop of a society, into which he doesn't fit. In some tales, like Dostoevsky's *Notes from Underground,* the antihero is a round peg in a square hole. But in nature fiction the outcast is often a walking piece of the wilderness, beyond domestication—someone who balks at being hobbled by society's laws or limited to tribal settlements that quell the wildest and therefore most natural parts of his animal nature. He accepts his role in nature, among animal neighbors. Thus he has more in common with a bear or a tree than with another human, whose instincts and motives he may share but whose purpose eludes or disgusts him. Violence is seen as acceptable, even beautiful, when it occurs as part of nature's routine dramas. Rarely do we find a full integration into society on the one hand and into the environment on the other. A character usually must choose between worlds, which means choosing between equally important parts of his heritage. Neither choice is bound to satisfy completely.

Alas, time goes only one direction in nature, and in the life of all organisms, from cauliflower to stand-up comedian. It has a beginning, a middle, and an end, in that order. By inventing stories, we stop that fateful trajectory, seize the second hands of all the clocks in the universe, and for golden moments nature becomes a realm we can revise. One of the things I find endearingly strange about human beings is our passion for reforming the world, not just so that we may survive with less toil and insecurity, but to reflect our innermost feelings. In the cosmic scheme of things, that's a remarkable thing for matter to do. Remarkable for upright apes, part of whose niche is creativity, but even more remarkable for organisms whose atoms were forged in an early chaos of the sun.

36

Seedpods decorate the garden from season to season. At first, they contain an Orient of riches. Giant pods have developed on the false indigo, and I love playing pendulum games with them. They look like green pea pods and make a wonderful thuddy rattle when shaken. Because they sit atop long stems and weigh more than the leaves, they move forcefully when you thwack them, swinging back and forth faster than the leafy arms of the plant, and even picking up a counterweight rhythm. Late in the season, when they turn a deep eggplant-purple, their weight drags them down to ankle level and it's noisy fun walking past them. Last year, I gathered every beautiful stem of the false indigo for arrangements, so I never saw the pods. But the plump purple and green pods are as beautiful as flowers.

The poppy forms a perfect round pepper shaker with tiny chambers that release one seed at a time, as if it were a pill-minder dispensing the day's drug. The columbine creates dry brown purses, open at the top, and I usually thwack them as I pass, scattering seeds higgledy-piggledy. I like to shake the pods, sprinkling hundreds of seeds on the ground beside each plant. I shake them not because it's efficient to, but just because I like the *shak-shak* sound of the seeds as they richochet around the pods and fly out. Some people harvest seeds more carefully and plant them with a clear method in ideal spots. I figure offspring will grow well where their parents do, but I also like the feel of batting seedpods and watching the chaotic panspermia that follows. Rose campion seedpods, rounder-hipped and darker, respond to idle slaps in the same way the columbine does.

Slapping seedpods can take many forms: the sly, as you pass, gosh-it-wasn't-really-me slap; the knockout punch right in the chops that rocks the seedpod without breaking the stem; the one-fingered flick that tends to shoot the seeds up in a fountain; the Three Stooges multiple-cheek-slap, which is a light flurry with an open hand; the teapot tip, which is more of a pour than a slap; or, a personal favorite, the catapult, in which you bend the stem back as far as possible without breaking it, and then let it spring up, flinging seeds in an arc. I suppose some might consider this unnecessary plant abuse, but I think of it as wholesome horticultural sex games. The plants need to procreate and I'm at least as considerate as a deer's intestines.

In autumn and winter, when pod-thwacking time has passed, the dried-up flowers and chattery seedpods of some plants look so pretty I leave them in the garden to add texture and shape during the winter months. Thistle forms spiny beige pods that are hard to handle but as otherworldly as sea creatures. Black-eyed Susan seed heads are dark and cone-shaped, like naked strawberries. Daisy seed heads always remind me of the vaudeville performers who spin plates atop long poles. Queen Anne's lace pods look like snowflakes when they're open and birds' nests when they're closed. Snap-snaps develop little brown purses with teeth atop long stems. Common mullein evolves tall, velvety candelabras, which people used to dip in tallow and hold as torches. When the cuplike flowers fade, seeds fill the cups, but tallow can fill them, too, and burn like tiny individual lamps. The hard pods of milkweed are crammed with silky parachutes, but even empty they look like beautiful silvery-gray mollusks. Asters form fluffy pods, sometimes delicately tinged with purple. Dame's rocket ends up looking like stick-figure people.

The cattails growing in the shallow marsh near the edge of my land are useful engineers. In time they'll dry out the marsh a bit by surrounding themselves with decaying matter and then creating soil. But I love their tall, velvety brown pokers, which some folk call "cigars." Inside each one, over a hundred thousand seeds are swaddled in soft stuffing, which will carry them across wind and water. In fall they look like old rags on stems, molting elephant seals, or moth-eaten couches left by the roadside. Surely pranksters have pulled out their stuffing. The usual pranksters are mice in winter and birds in spring, since the fluff makes a warm nest lining. Humans, too, have used cattail fluff for insulation, jamming it into their clothing on cold days.

In truth, all these pods are carcasses, mummified remains, the rotting remnants of youthful bloom. But I try not to think of them in that morbid way. Bland euphemisms like "interest" or "variety" usually disguise their fate, as in "seedpods create interest in the winter garden." A scene of carnage is what they create, but we're so easily bored that we crave any novelty in the bleakness of winter. We're so detached from the passion and purpose of a plant that we can find its dried-up forms beautiful. I'm guilty of that, too. I love growth, but I also love form, and few things rise to the architectural beauty of a plant responding to the pep and peril of the seasons.

37

The birds are growing restless, sleeping less, eating more, feeling hormones change, as they form long ribbons and wedges, migrating toward a distant summer. I'm always amazed by their calendar and weather sense. Somehow they track the seasons, and also the time of day, and know to wait for a favorable turn of weather before setting off. The sun is their compass, which they can locate even on cloudy days by finding its ultraviolet and polarized light. Although the hummingbirds have already left, I keep the feeder stocked with sugar water, in case migrants want a drink.

A red-tailed hawk wings overhead like a shard of stained glass sailing across the sky. It is a pair of binoculars with wings, its vision eight times sharper than our own. The hawk's eye has an extra set of muscles to change focus, photoreceptors packed tight and deep for tremendous magnification. It's so large that it can't even move in the socket—the hawk has to turn its whole head to follow a moving object.

Like everyone else who watched Alfred Hitchcock's *The Birds,* I found the clouds of birds menacing as they coated telephone wires and blanketed rooftops. Of course, they were supernatural and ready to stage piercing warfare with Tippi Hedren, but a truth becomes apparent in this season: Hitchcock didn't need to fake the shots, since that's what birds do as they prepare for migration. He could easily have filmed the scenes in New England or northern California on an average fall day, when hundreds of sparrows, starlings, or finches will perch along a stretch of telephone wire, hang on to poles, and wait on nearby rooftops for the seemingly magical moment when they all lift off together in a huge black cloud to do maneuvers in the air. It's wonderful to watch birds flock, then disappear from view as they bank edgewise, only to appear again with the next swooping turn. Formation flying isn't easy, and in the fall birds practice gathering with their kind, flocking, and group flight before they actually set off for the south, swirling like huge plumes of soot over the rooftops.

When food starts growing scarce, birds must move on. So must whales, fish, locusts, wildebeest, and many other animals. Most humans used to migrate with the same urgency, and in some parts of the world, nomads still do. But when birds blanket the sky in wave upon wave of beating wings, you can see calendar pages turning. Like ancient mariners, they steer by the stars and landmarks, and by a magnetic compass as well. Homing pigeons have deposits of magnetite in their brains that respond to the earth's changing fields. Floating above the earth, the pigeons are compasses, their beaks pointers. Some birds use scent, following the fragrance of familiar marshes, oceans, and meadows. Some hear natural sounds beyond the range of our own hearing—perhaps waves crashing against coastal rocks or the prevailing winds blowing through a high mountain pass—and still others use the planet's changing barometric pressures and gravitational forces to navigate.

"Birds migrate," we're told, as if it were one simple act, like catching a train. We assume they know how, when, and where, the god of instinct telling them: "Fly east, go now, I'll tell you when to stop." But in truth birds are adroit navigators with a highly detailed map sense. Combining techniques, they can orient themselves neatly, and they're not limited to just one strategy. Nature seems especially good at backups. Consider the ornamental lily, which in autumn forms shiny black snail-like bulbils along its stalk. The easiest way for the lily to reproduce is through scaling, when sides of the bulb simply pull away and start their own plants. But if that fails, the less reliable bulbils can fall to the ground and sprout, and if that fails there are always the seeds, which take a long while to produce an adult lily. But one way or another, the lily *will* reproduce. Birds, too, are well equipped for adversity. By day, they can read the sun, whether it's visible or invisible, by night the moon and stars; and they can use sounds, smells, gravity, barometic pressure, landmarks. If all else fails, elders can guide the young in flocks, so with one thing or another, they survive what seem almost impossible journeys. They know to wait for the best weather, how to catch a ride on a cold front, where the winds will speed them along, how to switch to the trade winds, like changing subway lines. They have no choice really. They follow the herds, just as our ancestors did. *Their* herds happen to be insects and other small prey, which vanish in winter. Ducks and geese would starve if they waited while the ponds and lakes froze over. In the tropics, food abounds, but getting there takes a heroic journey. In September, radar screens track as many as 12 million thrushes, orioles, tanagers, vireos, warblers, and a host of other songbirds. In October, the great exodus continues, with juncos and sparrows and several kinds of blackbirds. Then the hawks and other raptors join the great pilgrimage. Seventy-five percent of all the birds in my yard (and in the Northeast, for that matter) will disappear. A tiny warbler may fly from North to South America in only eighty hours. The Arctic tern sets the world distance record, migrating all the way from the Arctic Circle to Antarctica and back each year, a round-trip of roughly 22,000 miles. I've seen terns arriving in Antarctica, not looking spent, as one might expect, but feeding deliriously in the cold, rich waters. But to me the best trick of all is the one I'm afraid is a myth—hummingbirds hitching rides in the feathers of migrating geese and swans. Wouldn't the swans mind? What would a hobo hummingbird feel like? Sure would be cozy for the hummingbird. Or perhaps that's my all-too-human bias, when I imagine flying on a bed of swan's down.

Two seagulls wing raucously overhead, and I laugh. They seem so unlikely in upstate New York, far from the ocean, on a lake without a shore. But flocks of seagulls nest inland by large lakes. As scavengers, seagulls adapt quickly to city life, picking among the litter and learning to avoid glass windows and bright lights. They're not lured by the lake but by what the lake lures—migrating birds, many of which become confused by building lights and reflecting windows. About 100 million birds a year die from colliding with buildings. In Toronto, seagulls take an active part in that carnage by herding migrating birds straight at the buildings so that they'll crash and fall. Then the seagulls eat them. They kill the most during spring and autumn migrations, when many birds fly low at night. How resourceful of the seagulls to use buildings as hunting tools. It's fun discovering that some things we think typically human—being warm-blooded, using tools, having a language, etc.—really aren't unique to us after all. As I was saying earlier, some of our behaviors are shared by plants, and many more are shared by animals, and for many of the same reasons, whether their brains are big or small, or even if they have no brain at all.

38

A hard day for gardeners—the first serious threat of killing frost tonight. Twenty-mile-an-hour winds, gusting to forty miles an hour, make it impossible to go biking, despite the brilliant sun and temperatures of 50 degrees. The winds also mean I won't be able to stabilize sheets over vulnerable beds. The strong gusts would make them all take flight and the yard would look like a fully rigged schooner racing across the neighborhood.

Sitting in the sun, I enjoy the harlequin beauty of the yard. All the gaudy zinnias, thick with new buds, the scarlet geraniums and the sea of confetti-colored tickseed and impatiens. And, of course, the roses, my beautiful roses. Five heavy blood-red blossoms of Don Juan dangle from leaf swags hanging between two trellises. Time to appreciate them one last day before they vanish. All That Jazz has two huge orange-red blossoms atop its highest spike. White phlox still blooms among the pink mums and purple asters. The coleuses, if anything, are mounding larger. It will break my heart to see them wither. The violent winds, if they continue, might just save the day.

On clear nights, cold air settles on the ground, but winds can mix warm high air into the foggy moors below. That's why growers in Florida set out fans and propellers on frosty nights, to stir up the atmosphere and bring down a touch of warmth.

In the afternoon, I plant more bulbs—everything from crocus to giant alliums and fritillarias. The large crown imperial bulbs smell distinctly skunky, mainly to ward off moles. There is an appealing rhythm to planting them: plunge a six-inch shovel into the soil, press it forward to pry open a hole, toss in a handful of fertilizer, settle the bulb securely in the hole, remove the shovel, smooth over the soil. Smaller bulbs go in with smaller tools, sometimes in a wide shared hole.

Then the winds die down, and for hours I cover plants with sheets, pillowcases, garbage bags, and upside-down flowerpots. I staple sheets or tie them to stakes, then secure them with rocks. I protect the annuals and also any roses with buds. Now the yard looks all set for Halloween, with pink, black, and white ghouls fluttering in the wind. One year I should paint faces on them. My plan is to keep everything covered for three days and uncover them on Monday, when a week of warm weather is due.

As the skies clear to a sunny blue, and the weather forecasts sweeten up, I think about the covering and uncovering of the yard. At noon, I remove most of the garden's wraps, so that the mummified flowers can breathe a little easier and drink in some sun. The weather forecasters are calling for clouds to roll in late, with an overnight low of around 40 degrees. As I pull sheets off the roses, I find some broken branches and buds. Carefully I tie sprained stems, and collar sagging rosebuds with green jute. As the day progresses, the sky becomes glass blue; it is like living in the bottom of a paperweight. A bike ride in the afternoon takes me past a large flock of goldfinches scouting their next banquet of thistle seed. When night falls, a full harvest moon hangs low in the sky, bloated, touchably close. If I let my eyes blur a little and my mind roam, I can see its three dimensions and appreciate with renewed wonder that it is a large, round, planetary object, another world circling the earth. By 10 p.m. the yard temperatures begin to drop and all the weather forecasts change. Clouds are stalled far from town, the night will be perilously cold. I guessed wrong when I removed the garden's wraps. In darkness and chill, carrying a flashlight, garbage bags, armfuls of sheets, I spend an hour re-covering many of the plants. This time, I simply drape an open plastic bag over individual rosebuds I think might be at risk. Chrys and Bill believe it is the frost itself that kills the buds when it settles on them. Thus a top covering may protect them. Persis believes it's the ambient temperature. Who is right? I'll get a better sense tomorrow when I inspect the plants, most of which are only loosely draped. I've completely covered the impatiens, coleus, and hibiscus with sheets because they're tropical plants, and I'll leave them that way for several nights. Tomorrow I might lift their cotton armor a bit and let them breathe. But what I really need are garden cloches large enough to enclose a few rosebushes or a small bed, spring-loaded cloches made of parachute silk that open automatically like camp tents. Traditional cloches are bell-shaped glass jars first used by French farmers in the 1800s to protect individual vegetables and seedlings from frost. In time, the term was extended to include head-hugging hats for women, and that's how most people now think of them, as a style of millinery. These days, garden cloches can be made from plastic or glass, recycled bottles, milk jugs, clear umbrellas, aluminum-and-screen "lanterns," joined plastic columns you fill with water (as it freezes, water gives off heat which warms the plant). But they're small. In the Northeast, we get at least one Indian summer, often two. Technically, Indian summer is warm weather after the first hard frost. For one night, a pop-up cloche could shelter a favorite bed of flowers. What usually happens is that the morning after a hard frost the garden looks as if it's been pruned with a cattle prod. This first frost is followed by two weeks of warmish days and clear skies, during which a sense of futility and cosmic injustice tortures every gardener I know. As Eleanor Perényi wisely observes in *Green Thoughts* (1981), "A killing frost devastates the heart as well as the garden."

39

A brilliant blue sky sets off the golden hickory leaves and the red burning bushes. The maples are reddening from sienna to scarlet. Although some roses have survived, they've slowed down. Others still offer buds, but they open with a long, slow strain. A gold-and-black monarch butterfly swoops overhead and disappears when it encounters a backdrop of yellow leaves. A few brave yellow daylilies are still flowering. Coneflowers still unfurl their pink pinwheels. The showy sedums and asters are thick with bloom. Pink anemones—windflowers we call them—open to face the sky.

Now that few roses bloom (because of the waning sun), I pick confetti-colored zinnias, which last a long time in bouquets. The snakeroots, covered in white tufts, are too pungent for indoors but elegant with an art deco profile, and the black-eyed Susans are beginning to mound into the sort of undulating drifts Gertrude Jekyll would have approved of. Giant caladiums and coleuses have begun to form their own canopy three feet above the teeming undergrowth. And the rosebud impatiens, though leggy and sparse, put on a colorful show of crumpled campaign ribbons.

Persis called on her cell phone at 9 a.m. and I met her on the path through the woods. We stood and watched steam rising from the pool, fuming among the tree limbs, ultimately forming a cloud.

"It looks like moss," I said. "Doesn't it conjure up images of steamy, languorous days in the old South, maybe along a Louisiana bayou?"

"I wonder why we associate steam with slowness?" Persis mused.

"Perhaps we imagine oppressively hot, humid days when we can barely move."

We watched the mist cloud in the trees for a few minutes, and then Persis continued down the path, toward campus about a mile away. I returned to change over my closet, putting away summer shorts, pulling out heavy socks. On the way indoors, I checked the thermometer at the small rose garden: 40 degrees. No problem.

Because I've agreed to visit a book group attended by some friends, I'm spending my morning reading the first eight incredibly gory sections of *The Iliad,* which is essentially a boy's adventure story. But just to be sporting, one might read it as a struggle between the warring factions of the mind, for whom base instincts, selfishness, and raw violence are stock-in-trade. The armies of the mind are constantly warring over trifles, myths; they are not civilized but brutal. All the adversaries are related. At times, mainly from loss and exhaustion, they are susceptible to truce. After that bloody morning, I return home to the peaceful sanctuary of the garden.

Today I bring in the last roses of the season. Most of them are in sprays of barely opened buds that may not bloom indoors, but I thought I'd bring them in anyway. On three, I wedge Q-tips under each of the outer petals, hoping to pry them open.

The temperature is supposed to plunge tonight, and I've finally decided to let the garden go the way of all petals. The roses have been budding and blooming in slow motion and will gradually enter suspended animation. Not enough sunlight. The zinnias are still blooming, though sparsely. The blankets of impatiens and coleus are growing leggy and spare. Gaudy, foil-bright, sun-spangled leaves have fallen to the ground, turning the yard into a leaf mosaic that camouflages many of the beds. The deer have begun to thicken their coats. So, too, the squirrels, which busily bury nuts from sunup to sundown.

In a green glass vase etched with wildflowers, the last bouquet of garden roses radiates scent and voluptuous color. There are several shades of red: orange-red All That Jazz; magenta-red The Dark Lady; The Squire, which is the color of congealed blood; the harlequin shades of Blanket rose (now pink-red, now lavender-red, now brown-red). The last Abraham Darby is gamely unfurling pink tissue-paper petals. The last of the golden Graham Thomas, huddled beneath a perky pink Carefree, completes the bouquet. When they fade, the season of roses will be at an end. And what a glorious season it has been! Over 1,500 roses, by my reckoning. Next year I'll enlarge the rose bed. It will be exciting to discover which roses survive the winter. I put in a lot of oranges this year—how will *they* fare?

It seems impossible still to be bringing in roses this late in the year, but I'm looking at a second vase, a fluted green one filled with two dozen red and pink rosebuds, a few of which are starting to open. Earlier in the season, I would never have dreamed of including buds, but an end-of-the-world mentality takes over in autumn. Some buds are still tightly wound, capped by long, graceful sepals. This is the time of the sepals. Willowy, crownlike, stretching beyond the tight body of the bud, they look like arms uplifted in worship or celebration. Some go off at odd angles. Each day new buds open into small perfect red or pink roses. Meanwhile the bouquet of buds is a slow-motion explosion.

Because we have to transplant Jerusalem artichokes, I taste one of the edible tubers, which look like potatoes and taste like crunchy soap. Some people use them in salads, but I don't care for the taste. We call them Jerusalem artichokes, though "Jerusalem" is simply a mishearing of the Italian word for sunflower, *girasole,* meaning, literally, "turns toward the sun." The more familiar globe artichoke is really a thistle whose flower hasn't opened yet.

When I look around the garden in late afternoon, I'm amazed to find two roses in perfect condition to bring in. How did they escape my notice? Both blossoms are on a bush of Carefree Wonder, an especially hardy rose. Funny how much I delight in their pink nuances. All summer I ignored them, regarding them mainly as a pink background, or roses I would pick to fill out a vase of more spectacular roses. Common as pig tracks, the Carefree Wonders never much caught my eye. Now I savor these two bright, delicate specimens. I love how the outside of each petal glows talcy white and the whole goblet of the flower falls open like loose petticoats.

40

Now that their fruit's fermented, the apple trees are ripe as a gin mill. I thought I'd leave the apples for the squirrels and deer—a kind of token, not a gardenful of 80 proof. But each day I watch birds come to get drunk, chirp like mad, and stagger in the grass till sundown when a sodden few lift off. The squirrels get so tipsy they sometimes run drunkenly into the road.

Shaken down by the transitory blackmail of prewinter, I'd like to relinquish all notions of a four-quartered year, in which months chug past like treads on a tank, each one separate and inviolable. For example, the ragged interface of fall and winter (roughly overlapping November) is an entire season in itself. Winds quibble, alternate, and drop. The air is full of the distinct hollow noise of seedpods clattering on the trees like tiny rattles. Rotting leaves mulch into a fragrant stew. And the meteorological surprise of each moment fills the psyche with exhilaration. The sensual experience is unique and indigenous to a hybrid season for which we have no name. It doesn't exist linguistically, this fall-winter polemic (I suppose we could call it "winfall"), and so for most people it doesn't exist in fact. I prefer thinking of nature as a free-flowing organism, not as a series of doors slamming shut. In a quartered year there are three months of rain and buds, three months of flowers and scorch, three months of leaf rot, and three months of snow.

It is November, the first month of winfall, that peculiar time when the air is cold without being chill and the wind sucks and howls in contradictory bursts, ramming a frozen myth into every bone. On a specimen day, one can watch the ivy on a brick wall shudder off all at once, like a horse flinching.

Birds still berry in the woods. Grass grows high enough to choke the lawn mower. And though I see houses clearly through the bare-armed trees, low brambles and brush still make a thatch wall. Soon, fearing snoopers, I'll draw the curtains at night, lest strangers watch us swimming between lit window frames, unshaven, bare-faced, amorous or bickering, letting our bellies out, picking our teeth, being homebodies.

This morning, we scouted the forest for dead wood so that when winter grows hoary and we yearn for our own private inferno in the grate, we'll be able to mosey through the woods by choice, not be driven by the greedy processes of a fire. At the edge of the yard, rounds of wood lie stacked: fuel for months. The decayed stumps look light as balsa. Sometimes I like to heave them until my shoulder jumps at the bone, and watch a four-foot hunk of tree fantail, plummet, spin.

I've still no idea what happened in these woods. I mean on the stone slab—icy, gray, and sacrificial-looking—we found next to a green plastic boat filled with dozens of rifle butts. Only the butts. A few feet away lay a mound exuberant with moss. No one will hazard a guess what lies below. The previous owner of my house was an avid bow hunter, and in the garage we found a table whose legs—to our horror—were rigamortised deer legs. I think he stalked the woods for deer, raccoon, wild turkey, and pheasant; field-dressed the game; made sacrifices over the stone slab; thwacked the carcasses to roast-size with a hatchet; and finally buried the entrails a few yards off. I quiver at the thought of a mound of offal! Paul says plant flowers there in a fructifying hum, but I can't. I'd always be thinking of organs and intestines. Did he perhaps sacrifice a rifle butt for each kill? Better that than a finger. At the Lascaux caves, one can see paintings signed by artisans with mutilated fingers. How slight a knuckle was compared to a larder full of roebucks or mammoths.

Indoors, I build a lusty fire in the grate, eat figs and tangerines, mull over a new poem. What a smorgasbord consciousness is: a feast of jumbled delicacies. As it burns to ash, the fire looks like a wasp's nest. Outside, a light snowfall is melting like rain onto the garden.

After lunch, I bring in a few remnant rosebuds (I doubt they'll open), a pale pink geranium (still blooming!), and a pink coneflower, and make a small arrangement for the kitchen. Then I think for some while about all the "arrangements" we make each day. For example, on yesterday's chilly bike ride, I saw a hawk perching in a tree, a great blue heron standing in the creek bed with its feathers fluffed up for warmth, and, in the distance, a man with a dogsled on wheels, preparing his team for winter treks as best he could.

For example, Susan Greene of Wildrun, the Wildlife Damage Management and Wildlife Reserve, arrives in her navy-blue truck at 2 p.m. A slender, buoyant woman with a long, grayish ponytail, she does a walk-round the house, and we chat a while about the habits of red squirrels. Then she leans a tall ladder against the roof and climbs up. At the base of the library's cupola, on either side, is a long heat-escape vent; and Susan can see at once where a red squirrel has been entering the house walls. She nails wire mesh over three of the openings, and puts a one-way door at the fourth. Next week she'll return and wire up that door. There's no squirrel in residence at the moment, but squirrels have several dens and the house seems to be the red's winter quarters. Mine, too.

"That should keep out all wildlife," Susan says. "Except honey-bees. And let me tell you, you don't want honeybees to get into the house and make a vertical hive in a wall. If that happens, the whole wall has to come down."

"Heavens! I'll listen for buzzing."

"Surprised you haven't had any bats," she says.

"Oh, I like bats," I assure her. "Bats are some of my favorite critters, but . . ."

She laughs. "*But* is right. Not in the house." Soon her work is done and she leaves for the next house being invaded by likable but unwelcome relatives.

41

"A Guest in the Garden"

As night falls and bats begin swooping overhead, catching insects in midair, I marvel at their graceful aerobatics. Thanks to a long forearm, a gliding membrane that stretches over bony fingers, and strong arm and chest muscles, they can flap like a bird. Up close, they don't seem birdlike at all, but furry and quaintly architectural, full of arches and spurs. Because their skeleton is visible through their wings, they look like heavily ribbed umbrellas or the mechanical drawings in Leonardo da Vinci's notebooks.

Bats have always fascinated me, and I'm grateful they devour hordes of insects each night. But I'm not as passionate about them as Karl Koopman was. Koopman was the most knowledgeable bat biologist of all time. If he were visiting my garden this evening, I'm sure he'd have many stories to tell about the little brown bats winging hungrily across the yard. And I'd ask him questions that have been bothering me for some while: Is it true that male Malaysian Dayak fruit bats have mammary glands and can nurse their young? In a pinch, could human males do the same thing? Is it true that female bats—like elephants, bottle-nosed dolphins, and humans—help other females give birth and act as experienced midwives? Bats are very secretive about birth and like to hide. But Koopman would have known and been able to cite chapter and verse.

Like Aristotle, he knew everything. When he died, a vast knowledge of biology died with him. Blessed with a photographic and encyclopedic memory beyond the ken of computers, he was a dynamic resource on which colleagues came to depend. How esoteric was his knowledge? My friend Merlin Tuttle, director of Bat

Conservation International, recalls an occasion when Koopman steered him to a paper unknown to other bat experts, published in Spanish, in a South American journal that had folded after the first issue. Out of respect and disbelief, colleagues sometimes amused themselves by trying to stump him with esoterica, but they always lost.

An enthusiast and scholar of the highest order, Koopman ultimately became curator emeritus at the American Museum of Natural History, but that stuffy title conveyed little of the extraordinary man. Among his many discoveries, he figured out how Caribbean, South American, and African bats got to their final locales, how they evolved, and why they're distributed as they are. Before him, few had studied the bats of the Caribbean. He reasoned that certain bat-rich islands were once part of the mainland where related bats thrive. When the glaciers had poured down from the poles, the ocean levels rose, creating separate island worlds, and the bats traveled among them. He did similar work on bats all over the world, yet only scratched the surface of the knowable. So many bats dwell in the unexplored canopies of rain forests. A few years ago bat-catching turned up twenty-five different species in the treetops; sixteen of them live only in the canopies. Some of them are rare, and most of them are mysterious.

Although Koopman traveled widely to study bats, his legendary knowledge was based mainly on research, not on seeing a great many animals in the wild. Send him an exotic bone and he could swiftly identify it down to the subspecies. But show him a creature still in its skin and he might just hesitate. Fieldwork posed special physical challenges for him, and led to adventures that are seldom told. Here's one trip that Merlin Tuttle recalls:

"I first met Karl in 1962 when we shared a tent for three months on the American Museum of Natural History Uruguayan expedition. As a little person, accustomed primarily to city life, he was understandably apprehensive but had assumed that he could quickly adjust. . . . Karl was a man of great generosity and warm disposition except when someone implied that he could not do something because of his physical handicaps. Then, he was formidable.

"On his first morning in camp, he insisted on accompanying me on a search for rare rodents in a thicket of saw grass and briars. By the time we returned his arms and face were badly scratched and cut and his eyes were swollen nearly shut from at least twenty wasp stings. Attempting to comfort him, I took him swimming, only to horrify him when a piranha took a nip out of my side just as he was entering the water. Through it all, Karl never lost his sense of humor."

Out of respect for his legendary intellectual accomplishments, obituaries don't mention the curious figure he cut. Thus, like some of the exotic bats he studied, much of Karl Koopman's private life was rumor. A mutual friend tells me Koopman was a man of ambiguous gender, born with a syndrome that left him unusually short, and possessed of a high feminine voice, fragile skin, delicate hands, and a waddling gait. As a child growing up in Honolulu, and then Los Angeles and New York, he must have been taunted mercilessly. That would have derailed most people. But by depriving him of more common avenues of challenge and fulfillment, his disability allowed him to focus on a field where he could excel. Just as it is both our privilege and panic to be mortal, it was both his horror and blessing to be handicapped. What led him to such a fascination with often-maligned creatures that live by night could fill an evening's imaginings. I'm not sure how or if I would ask.

These days when it's popular to search for the origins of one's misery and the thinnest line stands between cause and blame, I find it instructive to look at lives that began badly, perhaps with ridicule and shame, lives that might well have foundered, and ask, Why not? Often enough, such lives become triumphant. Complex and frustrating and imperfect, to be sure, but sometimes filled with genius, conquest, and high-octane wonder. "Resiliency" is the term psychologists most often use for this personality trait.

Karl Koopman was just such a man, someone of gasp-causing

extremes, a biologist brilliant beyond measure, furiously disciplined and self-willed, but also an eccentric man, emotional, volatile, as apt to cry as to curse a blue streak. Somehow he was both dispassionate and melodramatic. The plight of the vanishing rain forests filled him with tears. Dedicated, good-hearted, and hardworking, he inspired an enviable amount of praise and affection. An ace historian, with a special interest in battles of the Revolutionary War, he could be solemn one moment and playfully prankish the next. Even though he was extremely busy, he always found time for people, from the greenest graduate student to the most senior professor. Few matched his generosity of spirit. Yet he never married, and although he had many devoted friends and admirers, to the best of my knowledge he was never in a romantic relationship.

In the backyard, two apple trees are tolling with fruit. Unhampered by their gnarled and twisted limbs, they bloom each year into towers of white petals and then rich pomander-like apples. Over time, they've developed into trees of great stature. This morning, for example, they filled the hungry bellies of three deer, and the hungry soul of at least one human. It's an apparition I think Karl Koopman would have understood. Perhaps he would have enjoyed these lines from "Maud" by Alfred, Lord Tennyson. They're not meant to be funereal, but a sensual invitation. When I learned of Koopman's death, though, they took on a special mournfulness.

Come into the garden, Maud
For the black bat, night, has flown,
Come into the garden, Maud,
I am here at the gate alone;
And the woodbine spices are wafted abroad
And the musk of the rose is blown.

Winter

You know what it's like in the Yukon Wild
when it's sixty-nine below;
When the ice-worms wriggle their
purple heads through the crust of the
pale blue snow;
When the pine-trees crack like little guns
in the silence of the wood,
And the icicles hang down like tusks
under the parka hood;
When the stove-pipe smoke breaks sudden
off, and the sky is weirdly lit,
And the careless feel of a bit of steel burns
like a red-hot spit;
When the mercury is a frozen ball, and the
frost-fiend stalks to kill—
Well, it was just like that that day when I
set out to look for Bill.

—Robert Service

42

Now that the winter spider's spinning its white web in the trees, I like to drive through the countryside of rolling pasturelands and farms. Here and there horses stomp about achy, as metal shoes echo cold up each leg, and snow cakes underfoot. What thick coats the out-turned ponies grow. They like basking in the blunt winter sun, the one season when flies don't bother them. Paul hates winter ("Death, death, death"), can't abide the fall, sweats too hard in summer. "Spring's nice, but a trifle brief," he concedes. When the temperature hits zero, or a blizzard begins, he says: "Here's that goddamn winter you wanted. How do you like it *now?*" Truth is, he just hates weather. Me, I love it, and would welcome another dozen seasons. I don't think of winter as hardship but as another playground (I can go cross-country skiing) and another lens through which to see nature. There will be record snowfalls—there are every year—but nothing as severe as people get in Colorado, where 90 inches of snow can fall during a thirty-hour storm. Yellowstone National Park sometimes gets 315 inches of snow per year. And then of course there are regions that get serious snowfall—northern Canada, Alaska, Greenland, Siberia, and the poles, places where temperatures plunge to 60 degrees below zero. Extreme cold can produce something I've always wanted to see, "habitation fog," an entire town swathed in a cloud formed by the frozen exhalations of the people.

Here winter brings snow, wind, and cold, but also less sunlight each day, thanks to the slight tilt of the Earth, when rays strike at a shallow angle and the sun just doesn't feel as warm. What an odd sensation. It's still the huge molten giant overhead, and yet its touch is less fiery.

After heavy snowfall, a house becomes one's shell. At least snails can carry their shells with them, and in a sense we do that, too, with our cars. But first we have to dig escape routes for them. As I pry layers of snow off the driveway, my teeth chatter and muscles twist in an icy clench. The wind scours my face like a wire brush. Our species prevailed during an ice age when a cunning animal was needed. I don't feel very cunning today, just cold, but I'm mortally grateful for the snow. Without snow, we wouldn't have the quirky brains we do. Without snow, there would be no snowmelt, on which our great rivers depend, and also a third of the world's irrigation.

The squirrels have turned the yard into a snow maze. What began as tunnels crisscrossing the snow evolved into corridors when their roofs collapsed. Like mythic mazes, these offer monsters, too—rival squirrels—lurking in some of the blind alleyways. Occasionally I'll see squirrel heads popping up from the snow as two meet and tussle until one retreats and hightails it to the base of a tree, which it climbs at speed.

Sitting at the patio entrance to the maze, a plump squirrel with a silvery tag on her left ear pauses to wash her face with falling snowflakes. I've seen a squirrel do this simple thing—wash its face with snow—many times, and it still makes me smile. As does watching a raccoon wash its hands in a puddle. There's something intimately familiar about such gestures.

The raccoons are now sleeping safely in their dens. Do they dream? Squirrels slow down in winter, but they still prowl for food. When the snow is higher than they are—often the case in New York winters—they make direct tunnels from tree to tree. I've noticed that squirrels get their bearings by looking at something first from one perspective and then another, as if mapping it three-dimensionally. This becomes especially apparent when they're trying to break into a bird feeder. Once they discover a direct pouncing approach won't work, nor acrobatically clutching it while trying to vacuum up seeds, they sit and consider the problem for some while. I watched one squirrel size up the situation for nearly an hour, carefully viewing the feeder from one branch after another, from the ground, the roof, then scouting different avenues of attack along the tree, good angles from which it might hurl itself at the feeder, thereby scattering seeds on the ground. I'm always impressed by their spatial sense and instinctive geometry. Squirrels are the tiny surveyors of the yard. On the metal roof of the feeder, it calmly considered this safe-cracking problem, inspecting each corner and seam. One element it couldn't seem to figure out was why, the minute it put its weight on the bird-perch, a metal door snapped shut over the seeds. But I could be wrong—it may have figured that out and just couldn't do anything about it. I'm not sure the extent of a squirrel's abstract thought, and as we've been discovering of late, many animals can do a surprising amount of what we call "thinking." Consider parrots that can count objects, identify things by color and shape, tell if something is open or closed. The National Zoo in Washington, D.C., has a fascinating think tank that explores animal intelligence, tool use, language, and social interactions. Our genes differ from the chimpanzees' by only about 1 percent. Think about what features constitute that 1 percent difference. Now think about the other 99 percent we must share.

This smart squirrel is a favorite female I tagged five years ago. She looks healthy and I'm delighted she has survived, since the average lifespan of a squirrel in the wild is only one year. Watching her as snowflakes fall big as bees, I remember the adventure of that tagging day. I've tagged several endangered animals in the wild—golden lion tamarins (orange, kabuki-faced monkeys with manes), monarch butterflies, and Hawaiian monk seals—with biologists from the National Zoo, the Los Angeles Museum, and the Monk Seal Project—but never a neighborhood animal in my garage. To mark animals for identification, biologists most often use Lady Clairol, bleaching numbers into the fur. We dyed the tails of the golden lion tamarins with ink, punched tags into the flippers of th seals, folded postage stamp stickers onto the monarchs' wi decided to check with a squirrel specialist, who had tagged hundreds of squirrels in the United States and Latin America, and abide by her counsel.

At 5:45 a.m. on tagging day, I woke anxiously and hurriedly showered, ate breakfast, and braided my hair, tying the bottom in a waterproof red ribbon. The ribbon brought back memories. Hummingbirds have sometimes attacked my red hair ribbons, mistaking them for swaying flowers and hoping for nectar. I've loved hearing them hover and whir near my ears. One doesn't need to leave home to go exploring, since we live surrounded by the marvelous, wade through it every day. It was a warm morning. The sun poured through the branches of the hickories where squirrel toys hung in a soft gold aura.

At 6:45 I set the traps, filling the back sections with walnuts, and placed a large heavy stone on top of each. Then I retreated to the house to watch. Within minutes, the morning platoon of squirrels arrived, hungry and curious. Some made do with the few peanuts and hazelnuts scattered by the windows, but the others, as I'd hoped, tried for the caged walnuts, first reaching through the mesh, then shouldering the cage, then systematically walking around the cage, then at last, tentatively, entering the cage. When one stepped on the metal trip-plate in front of the walnuts, the door snapped shut. Soon six gray squirrels were locked up tight. I checked my watch: 7 a.m.

The doorbell rang, and the rest of the squirrel wranglers arrived. Larry was first—the university veterinarian, who administered to a large array of research animals and pets, from camels to mole rats. He was a tall, slender man with large, strong hands, salty brown hair, and two earrings in his left ear. Deedra, a mammalogist whose home menagerie included a sloth and a kinkajou, strolled in with her long blond hair flowing loose. The last to arrive were Jaclyn, the authority on squirrels, who had driven up from New Jersey to lend a hand, and her husband, Greg, a contractor and fly fisherman. Jackie had long, dark hair and wore blue jeans and a black T-shirt with a large gray squirrel on the front. As familiar as gray squirrels are to homeowners, little was known about them, and we were all deeply curious.

"The squirrels are ready and waiting!" I said as I ushered everyone in. "Didn't need the peanut butter. The walnuts were irresistible." By now the wranglers had found a dozen squirrels munching nuts in Have-a-Heart traps.

"Wow, that's a mob," Larry said as we headed off to the garage to set up our equipment.

"There have been expeditions in Panama when we didn't catch any squirrels," Jackie added, "even though we were desperate to tag them. These are brave squirrels you've got."

"Brave and curious," I answered, "and well-fed."

Larry wheeled in two green oxygen tanks and an octopus of jars and hoses to deliver anesthesia. Deedra produced a white scale and a zippered net bag. Greg arranged a needle-nosed pliers, a wire cutter, and a tape measure on the long wooden bench that runs at chest level around the garage. Jackie sorted colored beads, lengths of stainless steel chain, and numbered ear tags. I fetched white sheets and thick yellow see-through plastic bags. When everyone was set with the equipment, I hurried out back to where the squirrels were waiting impatiently in traps, covered each one with a sheet, and carried them into the garage. Through the open door, I could see other squirrels exploring the remaining traps and knew we would soon have a quorum of squirrels to study. One captured squirrel was a red, which we turned loose. This was a gray squirrel study. Greg carried him out and opened the trapdoor, and the red rocketed away.

We put the first squirrel, cage and all, into a plastic bag, and Larry gathered the plastic at one end, inserted a hose, and began pumping in a mixture of 96 percent oxygen, 4 percent anesthesia. The bag was large and so was the squirrel, and with adrenaline pumping through its body, it took about five minutes for the drug to work. When at last we saw the squirrel sway and slump over,

Larry reached a hand into the cage and pulled it out by the scruff of the neck, balancing it on his knee, pressing it snug against his stomach, as I got ready to slip it into the net bag.

All at once the squirrel woke up and leapt straight out of Larry's hands, across the wall, down the floor, up a snow shovel handle, over a bookcase, around a rubber garbage can, and across the overhead door struts, then it ricocheted off a wall, dove under a pile of sheets, and ran between everyone's legs and frantic hands. In comparison, we moved in slow motion, as if under water. I tried to catch it with a butterfly net, Greg tried to tackle it, Larry swayed at the ready, like a soccer goalie, ready to leap in any direction, but it was Deedra who finally tossed a bunched sheet on top of it and lunged in with a leather-gloved hand to grab its neck scruff. I held its chest and bottom, and I could feel its rapid heart pounding against my bare hands. Its fur felt hot and soft. Larry settled a triangular oxygen mask over its nose and adjusted the gas so that the squirrel would finally sleep. While it drowsed, we turned it over to discover six large, dark, erect nipples with the fur brushed away from them. They were arranged like spots on a domino. Below them, large and prominent, was a clitoris one could easily mistake for a penis. Do female squirrels have orgasms? If not, what do they need such a large clitoris for? And why is it so unprotected?

"A lady," I said delightedly. "And with pups at home." She was a lactating female, and from the size and darkness of the nipples, an older, experienced mother. This was great news. Pups stay in the nest for as long as six weeks, but soon they'd journey round the yard with her. We placed her carefully on the bench, this small delegate from another species.

"Doesn't this remind you of alien abduction movies?" I observed. "Exotic creatures snatch up a human, put it on an operating table, fumble over its genitals, weigh and measure it, and tag it in some way for further study?" This otherworldly visitation by a higher species is what research biologists do all the time.

Her front feet had four toes with long, sharp claws, and springy pink pads that were much softer, paler, and thicker than I would have guessed. Her white tummy fur was silky and clean, and although she must have had some ticks and fleas, none flew up or crawled around for us to see. A healthy squirrel mother. Her tummy was white because of what's known as "countershading"—a form of camouflage. Any animal looking up at her from below would see white and perhaps confuse it with sky and clouds. Any animal looking down on her from above would see grayish brown tipped in white and perhaps confuse it with a forest floor. Like humans, squirrels let their mouths go slack when they sleep, so we could see her two huge bottom teeth, used like a can opener for piercing tough shells. Her black eyes bulged out far from her head, which is why squirrels can see over their shoulders, above their heads, and well below. If you live in the trees, danger can come from any direction. We have eyes on the front of the head, which means we evolved to be predators, not prey.

I fixed an ear tag in a special plier-like instrument and quickly snapped a tag into the left ear. We decided to tag females left and males right. Then Jackie wrapped a necklace of pink beads around her neck, slipped one finger underneath to make sure she had enough room to preen, swallow, and go about her high-altitude trapeze-artist ways without catching herself on a twig. It felt too loose, so we clipped off four links, then secured the necklace. Next we tucked her into the net bag, zipped it closed and placed her on the scale—she weighed a pound and a half.

"Wow! That's a big squirrel," Deedra said, eyebrows rising. As she held the cage open, I quickly unzipped the bag and slid the squirrel inside the cage, which contained a few walnuts. Then we set it down in a quiet corner of the garage—the designated recovery room—and placed a sheet back over it, to let the squirrel come round without too many distractions. A loud nut-cracking sound broke the silence. Somewhere, in a cage under one of the sheets, a squirrel was making the best of its lot and eating. Then another started to eat.

"Well, I guess they're not too upset," Deedra said, laughing. "I'm amazed, but they obviously feel comfortable enough to eat."

"Ready for the next squirrel," Larry said invitingly, and Deedra handed him a cage with an even larger squirrel in it. The process began again, though this time we squirrel wranglers were handier and kept hold of the female well enough to sedate, ear-tag, neck-tag, and weigh her—all in 5 minutes 20 seconds. It may have been a world record. Judging by her ear and face markings, I knew she was the bold plump one I called Collops. We weighed her, and watched the pointer swing completely off the scale! She too was lactating, feeding pups at home. Another healthy strapping mother.

"I think you may be harboring a race of supersquirrels," Greg said with a fiendish tone in his voice.

Certainly they were well fed for a year during the gray squirrel study, and then slowly weaned and allowed to return to their private ways. *National Geographic* had sent a photographer to follow the progress of pups inside a tree nest, and also record some of the more daring aerial leaps.

Now, five years later, three of the tagged squirrels still survive, browsing wild strawberries when I'm gardening, or drinking from the trough I put out for wildlife.

Finished with her washing-up, Collops turns on her haunches and scampers down one of the alleyways. If she had seen me, she might have begged for a handout, as she sometimes does, waiting patiently while I get her a slice of organic whole wheat bread. She always eats the crust first, turning the slice in her paws. (Kangaroos in the St. Louis Zoo hold and eat their 4 p.m. slices of Wonderbread in exactly the same way, crust first.) I don't know where Collops is headed, but I hope there's a warm nest and family at the other end and wish I could assure her that it will soon be spring, when the snow maze melts, flowers appear with their fragrant morsels, and the living larder of nature will be hers once more.

43

One afternoon, I drive to a hilltop home set on eighteen acres of wilderness, including a pond. There's something special about a pond high atop a hill. It reflects the fierce rootedness of tall grasses, the resiliency of wildflowers, but also the dome of the sky filled with an inexhaustible parade of clouds. At night, it seems to hold the tumbling well of space, the greater darkness between the reeds where moonlight skitters, and stars in a wilderness of stars. In the middle of the property stands a large, modern house with soaring ceilings, wide hallways, and airy rooms. Everything about the building speaks of open-endedness. Huge windows lead to the garden, pond, city overlook, and sky beyond. An easy step between worlds.

This was the first residential hospice in New York state, a refuge where the dying can spend their final days in dignity and grace; visited by family and friends; tended by physicians, nurses, and social workers; cared for by a militia of compassionate souls. Residents pay for their stay, but no one is turned away. Our community has kept Hospicare's doors open, though the residence wing had to close briefly and fund-raising is a constant challenge.

The building plans call for gardens, a comforting landscape that will include a chapel garden, butterfly garden, bird habitat, spiral overlook, small pond with waterfall, gazebo by the lake, and many interflowing glades and islands of flowers. But there simply isn't enough money to finish the project. Creative solutions are needed. To raise a small portion of the money, I proposed an all-city birdhouse-making competition, followed by an auction.

Birdhouses excite me in a special way. They fit spiritually and metaphorically, because birds are cheerful symbols of nesting and nurturing; they're melodic, colorful, and free-flying. Perhaps that's why, in a medieval work of the Venerable Bede's, life is depicted as a beautiful bird that briefly wings through a banquet hall. A young man who recently spent his last days at the hospice kept a journal in which he recorded daily delight in watching birds from his window, and there are many to watch. Ducks and herons visit, as do swallows, robins, bluejays, cardinals, crows, hummingbirds; also dragonflies, moths, butterflies, bats, and many, many other winged creatures.

When the gardens are finished, residents will be able to meander (on foot or in wheelchair) along its pathways, through nature's arbors, flanked by flowers and wildlife, to a gazebo beside the pond. Family and friends can also stroll there, picnic, and bring their pets—and hospice workers will get a welcome respite. There will be a meditation walk and many places along the path to pause, enjoy, and reflect. Visiting children can play, building snowmen in winter, stalking frogs and beetles in summer, jumping in leaf piles in the fall. There will be a butterfly garden. Whether residents physically leave their rooms or not, it will offer them the comforting embrace of nature. They will be able to watch the changing pageant of the meadow, pond, and sky. The antics of the birds will intrigue and amuse them. There will be a chapel in a glade. Hospice workers will be able to wheel residents along the pathways. I imagine the gardens will often be used for meditation.

Strolling in the sunshine each day, I sometimes find myself concentrating on the birdhouse competition, playing with possibilities. Some birdhouses might stay at Hospicare and become part of a small permanent forest of birdhouses. The ornithology lab feels strongly that the houses should be usable and plain, or, if paint is absolutely necessary, then water-based only, only on the outside, preferably in light colors. Would such restrictions limit the imagination and take some of the fun out of the contest? How can we leap that seeming hurdle? Maybe if there are several categories, some decorative, some useful, all with appealing names: for example: Simple Gifts, Shaker-Style; Colorful Quarters; Flights of Fancy.

But must simplicity really be so limiting? Perhaps I need to imagine simplicity in more elaborate ways. A simple house could be a gabled box with a slanting roof. Or a hollowed-out log could be topped with a nearly flat bark roof. It would need to slope a little to drain the rainwater. If the bark were rough enough, birds could get a good toe-hold and they wouldn't need a perching post. Simple hole-fronted houses might attract woodland birds like nuthatches, and if the hole were small enough it would keep out prowling sparrows. Simple houses with a wide rectangular letter-box opening on the front would appeal to robins and wrens, which like to keep an eye on the yard while nesting. But why think small? Simple boxes can be two-storied, like a chalet, with a feeding table underneath as a sort of porch. Bad idea. The feeding table might attract other birds, which would scare the nestlings. Now purple martins, they'd be happy with a condominium of ante-bellum stature. . . .

As we each become obsessed with the idea, the birdhouse caper, as I think of it, takes on the urgency of a mission and the fun of a game. We're swept up in a delightful frenzy. People keep hatching new ideas. One day, the local ornithology lab is roped into doing a bird inventory on the site and contributing design specifications. The next day, publishers are persuaded to donate bird books as prizes. Of course, they have to be wrapped in bird-design wrapping paper and tucked into bird-design tote bags. Another day, a painter is drafting birdhouse awards certificates. Then the bird-theme music must be selected—"Rockin' Robin"? Something by Messiaen? A committee member's fax proclaimes "Birdhouse Central" as its return address. Surely we need the city's most successful realtor to help appraise the birdhouses. Before the caper is finished, we have involved the whole community. A painter is engaged in a creative fugue, dreaming up birdhouse award ribbons and certificates; landscape designer Paula is imagining a spiral garden that curves like a chambered nautilus. I hope we are providing some of our neighbors with opportunities for creative play, as they dream up and build their birdhouses, but I also think about how precious the garden may be to the hospice's residents. Illness has a knack for narrowing one's focus. If someone has only a short time left, making contact with nature in an intensely spiritual and visionary way may offer special pleasures.

What wonderful creative mischief people get up to! We end up with fifty-nine birdhouses to sell. Most are ingeniously designed, colorfully painted, and usable by birds of all size and stripe. Some are elaborately decorative (to the point of having electric lights) and meant to be used indoors. Some are astonishing feats of woodworking, lovingly made by professional architects, contractors, and artists. Children build birdboxes of glitter and colored feathers. The youngest entrant, a four-year-old, has built a feathered bird tower that's taller than he is. A special favorite of mine, "The Oldest Birdhouse in the World," has walls paved with Devonian fossils collected along the lakeshore. Two purple martin houses are large, architectural Victorian buildings whose side panels open for cleaning (or playing dollhouse). The judges have a tough time—but a lot of fun—choosing winners, since so many are imaginative, painstakingly built, beautiful, and (often) hilarious. Essentially, we use the Westminster Dog Show rules.

Best of Show

1st prize ($500) to Dan Krall for No. 5, his architecturally sublime, 3-foot-by-3-foot copper-roofed Victorian Condo

2nd prize ($250) to Stephen Merwin for No. 19, his 7-foot-tall, spiky rocket tree with nest

3rd prize ($100) to Rachel Dickinson for No. 26, her pinecone-shingled chalet with back-door trellis

Best of Type:

Simple Gifts, Shaker-Style ($50) to Louise Adie for No. 31, her distressed-wood and clothes-hook bungalow

Flights of Fancy ($50) to Graham O. for No. 30, her lady gourd-bird with woven rattan wings and long, spindly legs

Colorful Quarters ($50 dinner for two at Tre Stelle restaurant) to Janet Locke for No. 35, her boldly painted Gaudi-esque sprung-star house

Best Children's

($50) to Cathy, Alize, Jamie, and Chloe of IACC After-School Wildlife Program for No. 23, their tower of colorful feathers with pipe-cleaner handle

Honorable Mention

Nature books to a dozen or so others, whose entries are so remarkable we feel they should be recognized

On auction day, storm clouds tower as we display the birdhouses indoors, in Hospicare's basement, on six long tables which we've decorated with flower arrangements (in swan, crane, and nest vases), pinecones, apples, acorns, evergreen swags, and toy birds. The room quickly fills, and as a local bluegrass band plays bird-titled songs, guests eat bird-shaped cookies and drink strawberry punch. The birdhouse committee members can be identified by copper birdhouse pins. What begins in a sociable mood soon grows playful, good-humored, and slightly wild. True, it is a small-town audience gathered to support a good cause, but the birdhouses are also collectibles, and some are funny. Every last birdhouse sells, including all of the children's feathered mysteries. One woman has bought the heaviest birdhouse, and also the most delicate, both of which will have to be delivered. Bidding is especially intense on a few houses, including the fossil house, a Shaker-style bungalow with dovetailed joints and exquisite attention to details, and two houses with meditations calligraphied around them as part of the design.

People leave laughing and gabbing, clutching new birdhouses. We have raised thousands of dollars for Hospicare's gardens, and also introduced a few more people to the residence and its work, so by day's end, the birdhouse committee is raw adrenaline and smiles. "See you next year," one bidder calls as she leaves, stacked high with birdhouses. A great outcome. It has taken much work, much planning and worry. On the other hand, it has also provided camaraderie and fun, taught us all many things, from how birds live to the hard work of staging nonprofit events, and helped finance gardens we hope might soothe.

44

"Dangerously cold," the weather reports said. "Don't go out!" For once, the radio's emergency test siren was for real. "This is not a test!" it warned, and then repeated the warning a few times because it had cried wolf ever since most of its listeners were in elementary school.

By now the true hibernators, such as woodchucks and bears, have settled in for a long, deep sleep. Chipmunks, on the other hand, are spasmodic hibernators and will wake up from time to time to go to their pantries for a snack. The great migrating flocks of birds have arrived at their winter roosts. Butterflies, too. Their migration seems even more miraculous. One brutally cold December day a few years ago, I flew to southern California, where I found colored lights flashing next to bathing suits in store windows and children stalking sweaty Santas through toys and tinsel. I had come to tag thousands of monarchs in nearly fifty overwintering sites along the coast. A hundred million monarchs migrate each year, and some fly as far as 4,000 miles, as high as 10,000 feet, just to find a warm spot to winter en masse. Many butterflies only live for one day, but overwintering monarchs can live for nine months (a hormone keeps them less active). Unfortunately, many of their favorite retreats are being obliterated by golf courses, condominiums, ranches, farms, and other idioms of progress.

A colleague and I drove north along the coast, enjoying the sounds of the Pacific Ocean bashing its cymbals on the rocks, the hoarse barks of the seals, the tooting of the gulls, the cheerful gab of roaming teenagers carrying surfboards and Frisbees along the beach. All the monarchs' sites were near the coast, and many were in eucalyptus glades, pungent with vapor, where long ropes of orange-and-black butterflies hung like Christmas garlands. Some monarchs perched right at the tip of evergreen branches. Then their wings would spring open with a clash of color, arch wide for the sun, quiver and flex, like bright ornaments. When dozens of them drenched a tree, their wings extended the angles of each leaf in half a dozen directions and gave the tree a trembling depth, a surrealistic wobble. It looked like a pop-up storybook. Overhead, the air thickened with orange gliders planing calmly back and forth. Sometimes the sun shone through their four-part wings as if they were stained-glass panels en route to a cathedral.

Surrounded by that drama, we sat on the ground, our nets full of butterflies, and carefully removed one at a time, checking its sex, health, and possible pregnancy, then applying a tiny postage stamp tag to one wing. Because it was a chilly day, we breathed warm air over their flight muscles, then tossed them aloft like tiny parachutists. Circling for a moment to gain altitude, a butterfly would climb to its comrades high in the trees and join the garland once again.

Hold a butterfly in your hand, and you will truly understand the word *flutter.* See one fly straight at you at eye level, its delicate wings beating vortices in the air, and you will feel for an instant that *you* are flying. Each butterfly is unique. It's hard to imagine a fat butterfly, but like humans they do have fat reserves, and some are fatter than others. They have various builds, diverse color patterns, uniquely shaped wings, and different black-and-white speckled bodies.

In a fantasia of butterflies, we spent idyllic days of observation and tagging, working together but speaking little. The heavy eucalyptus oils kept insects at bay, and few birds bothered to nest, so our days were startlingly quiet and serene. The eucalyptus vapors rocketed me back in memory to the Vick's VapoRub my mother used to apply to my chest when I was little, little enough to be newly amazed by the wonder of butterflies, living color on the wing.

That December lingers in my mind as a time of uncanny gentleness, something one doesn't always find in one's garden, of childhood sensations suddenly regained, of timelessness and a solid, geological sense of peace.

45

Nature is growing its own hellish garden under the house's cement slab, and the stench from the heating vent is ripe and nauseating. Although I've closed the register in my bathroom and covered it with a large coffee table book about love, the fumes make their way to the register in Paul's bathroom, and waft gaily whenever the furnace runs. This morning, my architect friend, Dan, stands by the register and samples the smell. Dan is a connoisseur of house smells, detective of scent, deciphering, intuiting, employing his best-informed guess to read the complex breath of a house. Is it ill? If so, where does the wound lie? Is it localized? Systemic? Minor? Gaping? We are midsized animals who detect partly by smell, though we've relegated that sense to the crudely bestial—not a part of our nature we like to acknowledge.

Long-legged and thin, Dan looks like someone he might draw on a set of blueprints, and his demeanor is as steady as a spirit-level. Driven by an aesthetic of line and shape, he responds to spaces in a way I've come to recognize but not share. For him, ghostly suspension bridges sweep across invisible wells in the air, anchoring the finite, implying the infinite. He quests to make them solid lines of truth. I've seen how a misplaced angle can rile him. At such times, his pain isn't simply professional, it's visceral; he gets twitchy, as if the order of all creation depended on a swift rebalancing of elements. In his universe, it does and, of course, precision calms him.

"This isn't an extreme smell," Dan says after a moment's thought. "It's fresh, the smell of sewage traveling successfully through the pipe. Sewage that's been pooling and standing for a while has a slightly different smell."

It's fascinating to watch him sleuth his way around a house problem, methodically acquiring information and then testing possibilities. He seems to imagine the house in part as a narrative. And although most people have a commitment to the familiar and thus resist change, his job in this house is to balance what stays against what must be surgically removed. Again, as often in life, it's a matter of what resists and what falls away.

He suggests that workmen pull up parts of the bathroom floor and slab, using special saws, not jackhammers. They'll sever the tiled vanity from the wall, locate where the bath drain and shower drain meet, and run a brand-new sewer out of the house northward instead of southward. The old sewer could be closed. He might even fill it with cement poured down the cleanout pipe. Looking out the bathroom window he sees the colored flags blowing like a regatta on the lawn.

"Boy, there are a lot of things to avoid out there—telephone, gas, electric, water—lots of noodles under the front lawn."

Out we go to look more carefully. He locates a possible route for the sewer. Back indoors he inspects the sink, the furnace room, the possible course of the pipes sealed invisibly in cement belowground. A narrative emerges, a possible explanation. When the bathroom was enlarged five years ago, workmen had to join several pipes. It may be, he speculates, that the joint has been gradually decaying since then, and perhaps a gasket has rotted or shifted, resulting in a crack in the pipe that allows gases to seep out. It's probably been happening gradually, but new holes in the heating ducts provide sudden access to the house. That means that rerouting the furnace and pouring cement into the old heating ducts should work.

If it doesn't, plan B involves sealing up the gypsum board walls in the furnace room. If that doesn't work, plan C requires digging a new sewer, running one line of it forward through the bathroom and a second line out from the kitchen, partly under the cement courtyard, to link up with the main line. Not a preferred plan, but doable.

Once we specify these details, oases in the desert of my confusion, I feel calmer. There are backups. We know what might need to be done and can avoid building anything that might block a potential sewer line. With that solved, we move on to the problem of the altered heating ducts. How to conceal them when they're running overhead? Do they stay exposed gray metal, as in a high-tech factory? Round? Dan feels that since the house's lines are all straight, the ducts should be square and hidden where possible. But how?

He decides on a scheme in which the ceiling in the hallway will be lowered and a wing will course along the top of the bookshelves in the living room, and then cling to an exposed rafter. His preference is for a bold architectural gesture. I'd prefer something altogether more modest and less overt. But I don't mind change in my environment. Paul, who is away and therefore not consulted, finds change unsettling. It disturbs him organically, bothers him as an idea and fills his gut with a strong sense of panic. When the repair first started, I showed him a set of blueprints, which he carefully studied. A lifelong master of English prose, he was a trained carpenter in his teens, yet he now looked at me blankly and asked: "What does it mean here, in my study, when it says 'remove wall'?"

I thought a moment, then answered: "It means you're going to Florida, dear."

Even these changes will be a trial for him, but he understands only too well that the house is sick and must grow in order to be healed.

"Which structure would you prefer?" Dan asks, showing me plans for two radically different schemes.

"Surprise me," I say. He looks faint.

Then the furnace man arrives, and he and Dan talk for some while about the how and when of the conversion. A contractor will prepare things for him, and the furnace man will arrive with a team on December 20.

How hard we strive to keep nature out of our houses. Not all of nature. Not roses and lilies and lavender. Only what we've designated as bad or disgusting nature—sewage, rot, decay, dirt, insects, bats, rodents. We don't like being invaded. We don't like feeling soiled. We don't like our efforts of cleansing and tidying to be nullified. We don't like our foundation to be undermined. And we fear all we can't see. What horrors lie on the rim of our perception? What devils lurk in the hellish underground beneath the house, the soil where larvae swarm and unknown creatures may be buried?

We cannot simplify and clean nature, of course, not external nature or our own bodies. But we pretend the world is manageable, at least our small homesteads. I understand this, but still I am driven to contain the chaos, to organize the forces of nature whenever I can in my home. I install a furnace. I turn on lights. I summon water in pipes. Would I prefer the poetry of a brook trickling over worn pebbles? Yes, but it's not practical. And we are such practical animals. For highly charged and often hysterical creatures, we are remarkably mundane. Do I want a faucet on the west wall of the house so that I can more easily water the daylilies? Probably so. It would be practical.

Tonight I will drain the pipes to the outside faucet on the east side of the house, as I do each year when I'm sure there's no more watering and the temperatures plunge below freezing. Yesterday, I brought home a bouquet of store-bought roses, and also planted eleven amaryllises. Organic fruit is scarce. Winter has begun, a time of rest and repair—the gardens rest, the house repairs. My bone-house, too. One hamstring is vexing me as it did early last winter after a summer of daily bike rides. Lying in the bay window, I survey the dozy gardens and try to shake the horrors of the day by repeating this mantra, *Rest and repair, rest and repair.*

46

It's snowing like gunshot. At noon, it was 60 degrees and ripening. Now the bench drips a thin white glaze, the quaking aspens keen and sway. As a steady sift of snow falls straight down and sticks, the wire fence begins to look crocheted. The wind poofs a snowdrift until it sprays flour. Powder snow, they call it, the skier's friend. To me, it just looks like small white grains, but I know it's really a blend of column- and plate-shaped crystals that prevent the snow from packing down. The shape of the crystal determines whether the snow will stick, pack down hard, or form large drifts, and there are many crystalline shapes, dendritic being the best-known one, the classic star snowflake design one finds on sweaters and Christmas ornaments. But snowflakes form many shapes which aren't always visible to the naked eye. Most people didn't know what snowflakes looked like until the nineteenth-century publication of *Cloud Crystals,* with sketches by "A Lady," who realized she could catch flakes on a black background and peer at them through a magnifying glass. Then, in 1931, Wilson Alwyn Bentley published an atlas of thousands of snowflakes he had photographed through a microscope, calling it simply *Snow Crystals.*

Depending on temperature, humidity, and wind, snow crystals can develop into stars, columns, plates, needles, asymmetricals, capped columns, hail, ice pellets, graupels, and strange combinations, such as "bullets," which are bunches of columns topped by pyramids, or "spatial dendrites," which are six-pointed stars (sometimes with what look like pine trees standing upright near the tip of each point). Columns are hollow crystals, spatial dendrites are three-dimensional crystals, and needles are solid crystals that easily produce permafrost and avalanches. The stars form in low clouds, the columns in high cirrus, and so on, but as they fall they hit other crystals, break apart, build new forms, and soon a flurry thickens into heavy snow. When I was little, my mother would turn off the lights in the house and turn on the garden light, then she and I would sit on the rug in the living room, admiring the ice-glazed bushes in her Japanese garden and watching snowflakes dance like dervishes under the porchlight, as flurry became blizzard, and I knew there was a good chance I'd be staying home from school and building a snow house.

Like glass, snow looks glittery and solid, and yet its molecules are in motion, it's continuously flowing. I like snow's odd quality of pouring over and around things without breaking up, so that it creates pockets of air, overhanging eaves, accidental igloos, where garden animals huddle to keep warm. I like how solidly snow packs, and how tiny flakes of it can bring a large city to a halt, when snow is nothing but water and air, mostly air. I like how well snow insulates, despite its essential coldness. I like that snow hollows offer insects and rodents lanes of travel. I like that running or cross-country skiing through snow requires chocolate as the ideal remedy for "bonking" (depleting carbohydrate reserves). I like the many names people have given snow. With each name I learn to see snow in a slightly different way. The Inuits and northern Indians, famous for a vocabulary of snow, have words for fluffy snow (*theh-ni-zee*), smoky snow (*siqoq*), falling snow (*anniu*), snow that collects on trees (*qali*), wind-whipped snow (*upsik*), crusty melt-freeze snow (*siqoqtoaq*), the bowl-shaped snow at the base of trees (*qamaniq*), fine smooth snow (*saluma roaq*), rough coarse snow (*natatgonaq*), deep snow requiring snowshoes (*det-thlok*), and even the place where wind has blown the snow away (*sich*), among many others.

Soon snow quilts the ground and talcs the trees, muffling sound, stifling scent. "Garden writers are fond of claiming that every season is as full of beauties and pleasures as any other," garden writer Charles Elliott notes. "This is of course nonsense." He goes on in the same delightfully snide vein:

> Watching an equinoctial gale toppling the *Verbena bonariensis* and sweeping away the asters, the second flush of penstemon blossoms, the cosmos that came on so late, and all the other autumn flowers; opening the curtains on a March morning to see the magnolia flattened under ten inches of fresh wet snow; peering out from the porch into the sixth dismal day of non-stop May rain—the heart does not merely sink, but bumps along the bottom.

For most people, winter seems to provide the most bumps. I like the contrast of hot and cold: the warm-blooded animals trekking across the snow, the heat of a furnace or fire keeping one snug through a blizzard, tropical plants such as amaryllis and orchids blooming on the windowsill while snow falls like volcanic ash beyond the panes of glass.

How ironic that trees get naked in winter and we pile on extra clothes. When the trees stand naked, you can see how hard they've struggled to bathe their leaves in sun. Winter is their desert. Water abounds as snow and ice, but if it's not liquid they can't drink it. Going without water for four months requires preparation; they need to pare down to minimums. Growth is expensive, it guzzles food and water, so they stop growing, and since the most growth happens in the leaves, the leaves fall. (Evergreens solve this problem by producing waxy, needle-like leaves.) A tree invests its remaining energy in building next season's seeds, blossoms, and fruit. Very much alive, trees just shift their attention in winter;

they turn inward, make subtle changes, become a silhouette of inscrutable branches.

The whole point of a tree is to colonize the air, where sunlight is plentiful. If it grew on the ground it would find stiff competition from many low plants. So in time trees evolved a woody trunk and branches strong enough to support all the life-giving leaves. It's easiest to appreciate this quest for light when you're walking in a dimly lit rain forest. High above in the leafy umbrella of treetops, life basks and thrives. A great many animals and plants live *only* in the forest canopy, including the three-toed sloth, which climbs down its tree once a week to shit at the base of it, thus fertilizing its tree home, and then quickly climbs back up before a predator pounces. In a rain forest, plants compete savagely for their inch of sunlight. They're always prospecting for sun, the source of energy and vigor, a form of sun worship the ancient Egyptians would have understood. I like how trees connect the earth and the sky, leading one's gaze from the physical to the invisible, the concrete to the abstract, the rock-obvious to the ether of faith.

On a really cold day, I sometimes fill the house with the smell of baking bread, its perfume clinging to draperies and hair. Few things feel as renegade as baking in winter and steeping oneself in rich aromas one associates with a living larder, the farm fields and sunny green gardens. If you're lucky enough to spend a morning eating oven-warm bread while bathing in a tub scented with lavender oil and reading a garden book, then seizing the winter day can be a sensory glut. I enjoy the powerful aromas of plant oils, lavender being a favorite. But if I had to choose a scent vital to people of many lands and eras, bread would win hands down.

Although in sixties slang it meant money, bread has always been more valuable than gold, and magnetically fragrant. I can smell baking bread now as I stroll about the house. It insinuates itself from the laundry room at one end to the garage at the other. No doubt the local bakeries are also busy creating fluffy loaves to feed the proletariat of our hunger: our desire for bread. The scent of bread, still warm and damp from an oven—a little sour perhaps, or with a slight caramel sweetness—can transport you on a caravan of memories. For warm bread may be delicious to eat, but it is also an appetizing idea.

One memory it conjures up is not necessarily of one's own childhood, but the so-called simpler life of our ancestors. (It wasn't really simpler, as any farmer knows.) People planted the seeds that grew into the grain they harvested, chaffed, and ground before preparing bread for the oven. Imagine being able to relish the full history of your food as you enjoy it: the grain crop rendered into a seductive taste.

Bread has a social history, too. During the past millennium, when bread became a staple people relished in forms as diverse as pumpernickel, challah, baguettes, or wafer-thin pancakes, daily life had to change. Producing a steady supply of bread meant agriculture, granaries, private property, settlements for mutual protection, the end of our meanderings. Bread implied the security of home, where the least you hoped and prayed for was your daily bread.

People continue to break bread together, just as they always have, to seal friendships or forge alliances (a *companion* is literally someone you eat bread with). "Let us break bread together on our knees" goes an old spiritual. We usually eat with our families, so it's easy to see how sharing bread would symbolically link an outsider to a family group. It's unfortunate, I suppose, that typhoid causes the skin to smell like bread; during epidemics one would receive mixed messages of comfort and fear while eating bread and nursing sick loved ones.

Most bread is a soft, fragrant, cloudlike thing under a sturdy crust, a food both robust and voluptuous. It oozes alluring scents powerful enough to distract one from work or play. In many languages, the word for bread is male or neuter (*le pain, das Brot*), which is surprising, because women have so often been associated with the sensuality of bread and the practicalities of baking. In fact, etymologically at least, *woman* is bread making. In ancient days, the word for lady—*hlaefdige*—meant "kneader of bread." In

Latin, she was the shaper of loaves, identified with the verb *fingere,* from which we get *feign, fiction, figment.* Both women and fiction stroll down the same etymological road, back to Indo-European *dhoigho-,* a wall made of clay or mud bricks. It is also the derivation of the word *paradise*. A woman kneads the bread of the family, shaping it, warming it, combining the ingredients of different personalities. From her doughy body, she bears them; with her stubborn hands she forms them; in the oven of her love, she binds them together. The mother is mortar. Her task is nothing less than the creation of paradise. Her motto might be something like "Live, love, and break bread."

Realtors understand the intense emotions unleashed by scents, and one of their favorite gambits is to have bread baking in a house being shown to a prospective buyer, especially in winter. No one can resist the smell of cooking grain. After all, for much of the millennium bread was either an urgent necessity or a romantic ideal, both summarized by scent. Wafting from the kitchen, clinging to the walls of house and mind, the aroma suggests family meals in a home where comfort reigns and all hungers may at last be met.

47

"A Guest in the Garden"

One day in the depths of winter, I try to remember the sense-drenching smell of Abraham Darby, one of my all-time favorite roses. What was it exactly? Candied lemon peel, apple, cinnamon, and chocolate, as I recall. Fruitier than the flower called Paradise, not as earthy as Purple Tiger, less lemony than Intrigue, sweatier than Playboy.

Smells are hard to capture in words, especially ones as complex as the individual scent cloud of a rose. Hard for me, not as hard for someone like Gertrude Jekyll (1843–1932), a clever, multigifted, disquieting woman who didn't hide her appreciations or her dislikes. Sociable and charming when she wanted to be, she was a widely known eccentric and sometime recluse. When it came to gardens, she felt everyone was entitled to her opinion.

An upper-middle-class British woman, Jekyll showed an early gift for art, which she studied avidly. But, as her eyesight quickly deteriorated, she was warned to give up her beloved painting and embroidery, so she turned her creativity to vistas she could manage: gardening, photography, crafts, and writing. "For I have sight both painful and inadequate," she writes in *A Gardening Credo,* "short sight of the severest kind and always progressive (my natural focus is two inches); but the little I have I try to make the most of, and often find that I have observed things that have escaped strong and long-sighted people." That's a colossal understatement. She had astonishing senses, which she developed and trained until she could hear rustling in the grass and know whether it was a bird, lizard, mouse, or snake. She could identify most birds by the sound of their flight. She could identify trees by the sound of the wind in their leaves, even as the seasons changed and the leaves dried or hardened.

> The Birches have a small, quick, high-pitched sound; so near that of falling rain that I am often deceived into thinking it really is rain, when it is only their own leaves hitting each other with a small rain-like patter. The voice of the Oak leaves is also rather high-pitched, though lower than that of the Birch. Chestnut leaves in a mild breeze sound much more deliberate; a sort of slow slither. . . . I confess to a distinct dislike to the noise of all the Poplars; feeling it to be painfully fussy, unrestful, and disturbing. On the other hand, how soothing and delightful is the murmur of Scotch Firs both near and far. And what pleasant muffled music is that of a wind-waved field of corn, and especially ripe barley.

In art school, she studied color theory in considerable detail and later applied it to gardening, paying special attention to the tricks that colors play when they're together, or in changing weather. This is one of my favorites among her color observations: "On some of those cold, cloudless days of March, when the sky is of a darker and more intensely blue colour than one may see at any other time of the year, and geese are grazing on the wide strips of green common, so frequent in my neighbourhood, I have often noticed how surprisingly blue is the north side of a white goose." That would be wonderful enough an observation but, apparently, she continued observing the goose all day, because she concludes: "If at three o'clock in the afternoon of such a day one stands facing north-west and also facing the goose, its side next one's right hand is bright blue and its other side is bright yellow, deepening to orange as the sun 'westers' and sinks, and shows through a greater depth of moisture-laden atmosphere." When I first read that passage I smiled, because it's a classic art student's exercise: pretend to be an ant crawling over an object, slowly observing every detail. I don't think fading eyesight is what leads to that sort of hair-by-hair curiosity about the world. She would have tuned her senses even if her eyesight reigned. One thing to keep in mind is that standing all day watching colors change on the feathers of geese would bore most people. Hers was a personality drawn to sensory minutiae. Twice a month she would choose a spot as a sort of song perch, sit quietly, absorbing the garden sensations, then write her regular article for *The Guardian.* In time she produced over a dozen books and 2,000 articles and notes, which, as Barbara T. Gates points out in *Kindred Nature* (1998),

> can offer insights into how this kind of writing transforms the natural into the aesthetic. When Jekyll recast the site of the garden into the symbolic structure of garden writing, she was performing a double translation. The garden is itself a representation of nature revised by culture, a situation in which an aestheticized nature is already literally in place. Garden writing sets out to further reenvision this nature-as-garden linguistically, a daunting task.

Jekyll brought her painter's eye to her own real garden, but also to the portable gardens she created in her books. An enthusiast of the Arts and Crafts movement, she loved beautifully designed handmade things, abhorred factory-made textiles and personalities. Her bossy and charming books taught British gardeners about flower arranging and garden design, with uncompromising theories of color, texture, breeding, and enjoyment. As Gates wisely notes, in Jekyll's notion of a well-made garden, "nature is not simply imitated but reinterpreted."

Jekyll was a tough critic of plants and people both, a woman of aesthetic absolutes, and heaven help a flower that didn't meet her standards. Perhaps the stem wasn't stiff enough for her liking, or the hue not distinctively pink or red enough. Those she dismissed as having "a bad and rank quality." Every plant must earn its keep in her garden, deserve its place. She was not the sort of gardener who never met a flower she didn't like. "I have discarded numbers of plants," she declared, "some because I thought them altogether unworthy, some because the colour of the flower displeased me, others because they threatened to become troublesome weeds, and others again because, though beautiful and desireable, they were very unhappy and home-sick in my dry soil, and it was quite evident that they were no plants for me."

Cottage gardeners and gentry friends alike learned what they should plant, organize, and enjoy, which greatly relieved the strains of subjectivity. Above all, she felt a garden should contain "pictures," which the "garden-artist" creates. Sometimes she described them as "living pictures," and although she doesn't mention it in her writings, I wonder if the craze for *tableaux vivants* didn't influence her. Certainly the novelty of the camera did. She enjoyed experimenting with photography in its infancy, photographing family and friends and the same kind of primroses I grow under my apple trees. She created fabulous gardens for her neighbors, but she also gave Britons the idea of the garden as metaphor, as intimate companion.

Claire G. Martin, author of *100 English Roses for the American Garden* (1997), reminisces: "My first experience with 'Gertrude

Jekyll' was back in 1987, when I planted her in my entry border garden. Between March and August, she grew to over 10 feet tall!" Martin means the showy pink rose, of course, whose fat buds open to big fragrant flowers. I don't have a Gertrude Jekyll rose in my garden, but I intend to remedy that oversight next year and plant her when it's safe, around Mother's Day. Not because of the day, just the time of last frost. Jekyll never married or had children. People like to comment on her being a "spinster," because of her supposed ugliness, but there are so many holes in that argument you could drain greens through it. Few values are as ephemeral as feminine beauty, which varies according to culture and time. In any case, as my mother likes to say: "There's a lid for every pot." She may well have had sweethearts and simply chose to keep them secret. Some garden lovers in her circle were gay, Vita Sackville-West being perhaps the best known, a woman poetically roused by gardens, who once wrote of old roses: "Rich they were, rich as a fig broken open, soft as a ripened peach, freckled as an apricot, coral as pomegranate, bloomy as a bunch of grapes."

Despite or because of her extreme myopia, Jekyll had a breathtaking gift for sensuality, equaled only by Helen Keller's. Blessed with a well-tuned ear and especially clever nose, she worked hard to educate them. Her eyes recorded the garden's subtlest changes in mood. But perhaps the greatest of all her gifts was her capacity for delight. As a young woman she traveled the world, but later she stood at the equator of her own small world, where the sun always stood high in the sky. She didn't take her garden for granted, even when it grew familiar. Plants pestered and pleased her, accepted her without hesitation or judgment, and got under her skin. That she could identify a rose by its smell says less about her expertise than about her willingness to be engulfed by a sensation and her openness to enchantment.

The woman I meet in her books has a great capacity for joy, someone who relishes even humdrum chores. "Weeding is a delightful occupation," she writes in *Wood and Garden* (1899), "especially after summer rain, when the roots come up clear and clean. One gets to know how many and various are the ways of weeds—as many almost as the moods of human creatures." I suspect her other moods were strong as well. But a book is a special mental locale an author is willing to share with readers, who are guests in her garden. Jekyll seems to have relished both gardening and writing about gardening, which is no surprise, since a book is like a garden. There is always something to tinker with and improve; it has seasons: research, writing, production, publication; it inspires ceremonies, rituals, struggles, and rewards. She was at ease in both worlds, the physical garden and the recorded one staked out with language. Writing about an event allows one to re-experience it in slow motion, whether it's a flower or an idea. Both flower and idea are dramas. They are at the atomic level, of course, where all is motion. But an idea is also a tremendous state of excitement, and a flower is a cottage industry. It's a good thing she wrote about her gardens, because now they are gone, or rather the soft tissue has vanished, leaving only the skeletal remains, the structures where her beloved plants grew. I lift the cover to her books as if they were gates opening onto antiquities, the lost gardens of Atlantis.

Although she ultimately chose a small circumference for her world, she mapped it thoroughly. Painting, photography, embroidery and beadwork, gardening, and architecture were just a few of her fascinations.What fun it would be to follow her around my garden. "Tell me what you hear," I would say. "Show me what you see." Maybe it wouldn't be polite to use her like a precision instrument, but I envy her deep attentiveness. She would love the cloud hips of the summer sky, the finches flocking like a shake of black pepper in fall, the milkweed flowers' strong sweet liqueur in spring. I wonder if she applied her famous horticultural sensitivity to the rest of her life. Could she walk blindfolded into a house and know the occupation of the owner, or whether its owner was male of female, by scent alone? Did she delight in the sound of creaking stairs, the slippery ropiness of kneaded dough? No one thought to ask. Surely she applied her utilitarian credo everywhere: "Is it worth having? Is it worth doing? These questions form a useful mental sieve," she says at one point, "through which to pass many matters in order to separate the husk from the grain."

The sieve again. Mathematician Henri Poincaré writes that his gift doesn't lie in hatching copious solutions to problems, but in being able to apply a mental sieve that would let only the best ones pass through. When plants photosynthesize, they send nutrients (a flood of sugars) down to the roots, moving them from one cell to the next through microscopic sieves.

Did she apply her discerning sieve to friends and loved ones? Despite her reportedly "abrasive" ways, she had loyal and devoted friends, who addressed her affectionately as "Bumps" or even "Aunt Bumps." Perhaps the richest of her friendships was with architect Edwin Lutyens, a working partnership that was to last forty years. They met at a tea party and, despite the thirty-year age difference, took to each other instantly, in time developing a unique style of "a Lutyens house with a Jekyll garden."

When she died, an unusual painting of worn, muddy boots became a symbol to her devotees. William Nicholson, who had painted her portrait, also painted a study of her gardening boots, as if to say: "Who could fill these shoes?" Over the years, she's had a searching influence on architects and gardeners alike; some of her phrases and ideas have become part of their everyday vocabulary.

I like the way her own garden fit her, body and soul. She was short and wide, and so were the garden gates, which she had made to her measure. She didn't allow men, especially young men, to follow her through them. A bulky woman, she was especially fond of small, delicate Fairy roses, which cascade like pink petticoats right up to hard frost, and I wonder what she'd make of the new red variety in my garden.

She's such a bossy-boots that I don't know if I would have survived hiring her. That her creation would be exquisite, harmonious, and picturesque goes without saying, but it would express her own rugged personality, not my more fanciful one. I think some of her rules may be whims justified by aesthetic theory. The gardens she designed for the Edwardian estates might contain a long bed laid out in a crescendo of one color with all its moods and tints, or an elaborate garden with many levels, paths, and outdoor rooms. Still, I do crave a winter garden right about now, and she understood that atavistic yearning. One enters her winter garden through a double arch of stone:

> Walled on all sides . . . it is absolutely sheltered. Four beds are filled with heaths, daphne, *Rhododendron praecox* and a few other plants. These beds, in company with the surrounding borders and the well-planted wall joints, show a full clothing of plants and a fair proportion of bloom from November to April . . . and a seat in a hooded recess is a veritable sun-trap.

A sun-trap. The very thing in winter. Which is not to say that the winter woods aren't beautiful, especially when fog clings to the trunks of some trees, and others wear gray-green pantaloons of lichen. The prickly holly leaves surge with dark green, and the quaking aspens along the driveway attest to the garden's skill at reinventing itself. Aspen trees die young—I've lost most of the grove over the past twenty-five years—but they're extremely fertile, launching waves of sticky white catkins each spring which froth across yard and street. Tree sperm. Because saplings crop up quickly as elders wane, the grove could replace itself entirely in fifty or sixty years. Sometimes aspens reproduce sexually, and I may have lost all of one gender, since male and female flowers grow on separate trees. Hard as I search, I can't find any aspen seedlings, so perhaps I'm left with a single-sex grove. If so, which ones survived, males or females? I can't tell just by looking, and will need to research the matter further. Of course, aspens can reproduce asexually, too. We're used to having two sexes, and thinking in terms of male and female, but plants have worked out nearly a dozen ways to reproduce. An aspen grove, for example, looks as if it's made up of separate trees, but the trees usually share a root system, so essentially they're one organism with many shoots, an organism that can grow to huge proportions, cross a meadow or climb a mountain. There's a 106-acre stand of aspens south of Salt Lake City, Utah, that's reputed to be the largest living organism on earth. Visitors may think they're hiking through an aspen forest, but it's all one tree, about 47,000 genetically identical stems rising from a common root system.

Anyway, these aspens (or aspen clones) don't mind having wet feet and live happily in the small floodplain alongside the driveway. Sunlight flashes off the tops of coinlike leaves, which make a delicious rustling sound as they tremble in the breeze. Aspens belong to the willow family, whose bark and leaves contain an analgesic similar to aspirin. But aspens are also great healers to the woodlands, because they're among the first trees to repopulate a devastated area, providing shade and protection as other saplings grow.

To Jekyll, aspens offer a color lesson. The eye searches for subtlety among winter's limited tones. "The white of the bark is here silvery-white and there milk-white, and sometimes shows the faintest tinge of rosy flush. Where the bark has not yet peeled, the stem is clouded and banded with delicate grey, and with the silver-green of lichen." She goes on to note the many darknesses of bark, too, especially among older trees. I'm sure she would have loved knowing the Inuits' vocabulary for snow.

The searching eye prowls the winter landscape, noting the posture of naked trees, learning the beauties of glaze and glare, stalking colors, appreciating how plants and animals float against the white canvas of snow. In winter one can see the forest *and* the trees, so it's a good time to remove branches threatening the house or dead trees that could fall without warning. Jekyll plucked trees as casually as eyebrow hairs to achieve a graceful line in her woods. I prefer to leave fallen trees in the woods for animals to enjoy and fungus to homestead. This doesn't lead to a tidy forest, but it does create horizontal and vertical lines, and a nice variety of shapes and heights to enjoy. Hefty tree trunks provide miniature ecosystems, which attract mammals, birds, amphibians, insects, fungi, algae, and even wildflowers.

Anyway, I think Jekyll would enjoy watching the tree surgeons who arrive at midday, wearing orange hard hats and looking a little like astronauts, to remove whole trees and heavy branches threatening the house. The raccoon warren, halfway up a colossal oak, makes that tree precarious, but I'm only worried about a high heavy branch that arches directly over the house and might well thunder down in a storm. Thus a brave young man is sixty feet up right now, in a rope harness you'd think would cut into his blue jeans. One rarely gets to see human primates swinging through the trees like our distant ancestors. Hugging the tree, working his way around the limbs, he is tentative and careful, but all primate. His moves are sure as a rock climber's, and he's superb at rope work on swaying limbs. While watching him from inside the garden room, I snack on a bowl of organic cherries. In summer, I pitted and froze small bags of them as a winter treat. Since I prefer organic fruit, I dine on what's in season, and in the winter I long for fresh cherries and grapes, fat blueberries and shining strawberries. Wise King Boethius advised that one shouldn't expect to find new wine at midwinter, but of course we do anyway, hankering for the fruits of summer.

The tree man has stopped to smoke a cigarette. Sixty feet in the air, he's slouching in the tree crux and harness, casual-like, with all the time in the world. He might be sprawling on a porch chair. Jekyll would advise and bully him, I'm sure, and probably direct me, as well. I doubt she could resist. The chains of perfectionism are heavy, as Oscar Wilde once said of marriage, adding "It takes two people to carry them . . . sometimes three."

48

If gardening is the slowest of the performing arts, its winter hiatus can feel painfully long. Surely the Earth has stalled in space. "The most serious charge that can be brought against New England is not Puritanism but February," Joseph Wood Krutch grumbles in *The Twelve Seasons* (1949), adding, "Spring is too far away to comfort even by anticipation, and winter long ago lost the charm of novelty."

One grows restless to be growing things. Gardens promise reincarnation. A garden is nature in miniature, under one's guard and manipulation, if not complete control. Watching nature grow, one connects with one's own growth, not simply from point A to point B, but the way one grows into an activity or idea. So, like hibernating animals, gardeners wait in suspense for spring. Nature is beautiful in winter, and we *need* to find it so, even if that may mean finding beauty in death, and spinning whatever fairy tale one requires to bridge the unbearable subtraction of flowers. Not everyone is sanguine about it. I enjoy Jamaica Kincaid's blunt declaration: "It is winter and so my garden does not exist; in its place are these mounds of white, the raised beds covered with snow, like a graveyard . . . I always take this personally, I think a frost is something someone is doing to me." It doesn't help that a newly frosted garden looks as if someone took a blowtorch to all the flowers. I try to make the best of it, to find pleasure hidden somewhere among the leavings. In winter, I search the garden for eye-catching phenomena, scraps of life, or remains that tell a story, but the experience of beauty is different in winter than it is in summer, just as reading a book is different from watching a movie. In summer the garden showers over you, and even sitting quite still you can relish its busy hive and waving colors. In winter, enjoying the garden requires more effort. I try to think of the beds as hibernating, not ruined, because I know they will return with gusto in a few months, and I don't want my senses to starve in the meantime. I also enjoy the novelty of seeing everything from a different perspective, the twigs glazed with ice, the snow glittering like diamond dust. In winter, I no longer respond to the garden's growth, but to its fetching phenomena, and I'm a pretty good beauty hound.

For example, this is the first time all year that the garden bares its bones, displays the skeleton on which its lusheries rely. Rose canes arc gracefully over the trellises, adding lengths of rich purple and green, while fat orange rosehips and smaller red ones button up the beds. Some of the most beautiful canes belong to the wild raspberries we call "black caps," whose plum-colored limbs blush with a white sheen. It's easier to locate the berry bushes in winter, when they're not hidden by brambles, and when I go cross-country skiing I make a mental note where to find their luscious fruit come spring.

The amaryllises have begun blooming like yachts along the windowsills. Two red geranium trees stand sentinel in the living room, cheering red, festooned with pom-pom blossoms, just as they were all summer. Six potted orchids add comets of lavender and yellow to the indoor show. Although my favorite orchids are lady's slippers, which I'm told are easy as grass to grow, I've never had luck with them. I swear, they're the teacup Yorkies of flowers—high-strung, adorable, and not good at managing their bladders. But I do have great success with moth orchids and dancing ladies and other common varieties, whose delicate long limbs and fleshy petals offer a wonderful reply to the trumpeting, thick-necked amaryllises. When you've found orchids growing sparsely in the rain forest, it's odd to see them potted in long military rows in a greenhouse, beyond whose glass canopy snow is flurrying. It's as if you've fallen into a fairy tale, and in a sense you have—the fairy tale of equatorial sun in northern winters. Nonetheless, I went orchid shopping this morning.

In the parking lot, I watched an elderly couple drive up in a shiny black car, drift into a parking stall, and turn off the ignition.

They climbed out slowly, stiffly, opening the doors in several stages and lowering one foot onto the ground, then holding the door for leverage as they pulled themselves upright and slipped from between the two banks of metal. The woman limped, her hip angled up with each step.

"Watch yourself now, Mother," the man said, with the habitual protectiveness of fifty, maybe sixty, years of marriage. His thin white hair, wan complexion, and cataract-fogged eyes all seemed shades of the same color. His dark suit hung loosely from his frame, and somehow he seemed not to be inside it, nothing did. Together they hobbled to the building with tolerant slowness, concentrating their efforts, saying nothing.

At almost the same time, ten yards away, a young couple pulled up in a rust-eaten blue Chevrolet, jumped out, slammed the doors, and met at the car's trunk, their hands finding one another as if by radar. A light breeze blew loose shoulder-length hair away from her face and four silver earrings. On her wrist was a tattoo of a rose. The man let her hand go and put his arm around her shoulder as they strode energetically, full of the day and each other.

Soon they caught up with, then passed, their elders. I've seen scenes like this one at other times, other places, throughout my life, which haven't startled me as much; I guess an emotional immunity takes over. Today I felt pulled from life's insensible rush. An alien and neutral light fell upon things. The full horror of mortality, that everyone will grow old and infirm and then cease to exist, chilled me with a sense of doom. Stricken, I watched until both couples disappeared into the building, the young man holding the door for the older two, who smiled and thanked him as they walked in, and his young woman, impatient, or unable to look at them, watching the cars pulling into the parking lot. Then she bounced into the store, and he followed, giving her a pat on her rump.

I knew I would have to wait until the sense of detachment passed, and I could return to errands, destinations, plans, then reset the machinery of denial. Evolution is no chump; if we saw to the end of the tunnel all the time there would be no going on, no babies, no cities, no hope. Hope is a fortifying confection, something to get one through the long hours after dawn. All roads led to the Rome of my despair. I knew exactly what was in store for me. It was unacceptable, and it was unchangeable. Was I the only person in that parking lot who had at that moment a vivid sense of her own mortality? I looked around: the cars, the people coming and going. I noticed, in passing, that my keys were in the ignition, and I couldn't remember whether I had just arrived or was ready to leave. Yes, now I understood how it worked, why one thought about such frightening things as death so rarely. It was like focusing on the up-close and the distant. If you concentrated on living, the human condition blurred, mercifully (otherwise you wouldn't be able to function). And if you changed focus and stared mortality straight on, "saw the fire," as Wordsworth said, then your other vision, the one that got the spoon from bowl to mouth and made dates and conceived and threaded needles came to a halt. If you dwelt on death for too long, your biological system would grind to a stop and you would die.

"Are you all right?" a voice called gently into my left ear. I turned my head to see a teenage boy in a white shirt and black bowtie, holding a few empty grocery boxes. I read conflict in his eyes; he wished to be solicitous but not intrusive. And, as if a paralyzing electric current abruptly stopped, I was able to smile and say, "Yes, I was just thinking," then turn the key in the ignition, put the car in gear, and drive away. By the time I crossed the intersection and traveled halfway down the block, I was already driving with habitual ease. We begin our lives as saltwater animals, suspended for nine months in a saline fluid in the womb. We grow. If we are very lucky, we grow old.

According to Pliny the Elder, places reveal in subtle ways the ceremonies and deeds that will flourish there. The art of deciphering those qualities he called "geomancy." The Earth is a garden in space, and we are some of its blooming life. How odd to be a sack of chemicals that can contemplate itself, and how much fun. A stone in my front courtyard, at the base of a hawthorn, is engraved with the words "*Carpe diem,*" seize the day. Next to it, a smaller stone reads: "Wonder." Lying together, they become two faces of the same thought.

49

My mother once visited the hanging gardens of Babylon, though, of course, she found them changed. According to legend, King Nebuchadnezzar built raised gardens in the heart of Babylon around 600 B.C. Babylon lay on the plains, and his wife was homesick for the mountains of her birth, so the king built her terraces, held aloft by stone arches, which were essentially raised parks complete with soil and trees.

Now they're bare ghost gardens, enclosed by vendors, and can be viewed on a postcard. All historic gardens live in the imagination. As do the gardens of childhood. Or the gardens of past loves. Heraclitus was right when he observed that we never step in the same stream twice. We breathe the same molecules of air, we till the same soil, but a garden left untended by laziness or death vanishes because, like a life, a garden is a cultivated space. My garden lives in this book, but when I am gone it will soon give way to another's aims, or perhaps become overgrown, or bulldozed for closer-fitting houses. I'm planning to bequeath the woods to the local Land Trust so that it will always stay a wild vest-pocket park. Meanwhile, I guide its growth, and it accompanies mine, which is the way of different generations of living things.

It's a fine day for reading the commerce of the yard. Bird tracks, squirrel tracks, rabbit tracks, vole tunnels, and deer droppings (what naturalists euphemistically refer to as "scat"). If you think about it—and I can't imagine why you would—in most mammals, the sphincter muscle pinches off poop it has expelled. Thus, if you look at the point on the scat, you can figure out which direction the animal was probably heading.

Winter means opalescent snow, cross-country skiing, and making maple syrup snow cones, but not the absence of birds, many of whom remain in the yard and make do with what food they can scrounge. Insects hibernating in tree bark become food for the chickadees, woodpeckers, and titmice. The finches, sparrows, and cardinals find shrub berries and flower seeds, and the crows and gulls eat most anything, especially on days when humans put out their garbage for collection. There are still owls and starlings and some ducks about, still some blue jays, doves, and dark-eyed juncos. You can tell the wind direction by watching them perch. They face into the wind with slicked-down feathers so that the cold can't needle its way in. On winter nights, the crows roost together in the tallest trees, forming eerie encampments of fifty or so raucous birds. One hears from time to time of a gigantic crow metropolis, usually in the South, where 200,000 birds may roost. Fifty to a hundred is tops in my neighborhood, and even when I don't see them, I often find their tracks in the snow (they drag a toe between steps), or hear their nerve-jangling call, which falls somewhere between a shriek and a bark. Their huge nests sway in the tops of the trees. Watching them, I'm reminded of the highest ship's lookout, called a "crow's nest," given that name in the days when seafarers kept crows high on a mast. If strong winds blew a ship off course and the captain lost sight of land, he would release the crows and follow them to shore.

A jet is skiing up the eastern sky, leaving white parallel clouds of ice behind it, before it melts away into the blue that is nothing. And in the north two thick contrails are meeting in a stately, full-bodied X. I wonder what pilot flew to the opportunity, chuckling as he or she crossed the abscissa. The two strokes are widening now like a chalk scrawl in blue and empty sky. Soon the air currents will disperse the X. Mark this space, the sky reads. Treasure buried here. Rest at this crossroads.

Low in the sky, a pair of cloud forceps grasps the sun, reaches deep into the green horizon, then sets the sun adrift. Another jet cuts a route through the loose X, and with one stroke opens it into a star, then bisects the sky so cleanly that, for a moment, it looks like the top half will rise up and the whole day blink. What fun the pilots are having! I remember what it felt like to play sky games.

When my dear friend Martin died in a crash, I continued flying for a while, but it was never the same. My cub pilot's innocence crashed with him somewhere between horror and possibility, and from then on the ecstasy I'd found in flying trailed a shadow of loss. But my years as a pilot were wonderful in many ways, and sometimes now, as I watch planes cavorting overhead, I miss the thrill of cloud-lapping, of a landing's controlled fall, of puzzling my way from one landmark to the next, of the plane folk who hang out at rural airports and compose a unique culture with its own humor and values. I wasn't a particularly talented pilot, never really mastered the different shapes of holding patterns or cared for flying blind on instruments (essential given the climate in my hometown). But I did love being a part of the sky, and floating above the Earth, wide-eyed, enjoying the changing seasons below. It's never the same being a passenger, as you are even if riding jump seat or in the copilot's seat. When tour planes were still allowed to swoop through the Grand Canyon, I rode as copilot in a twelve-seat high-wing plane, and I'll never forget the worried passengers behind us, screaming: "Don't let that girl touch anything!" "That *girl* is a pilot," the pilot snapped back at them, but it didn't matter. The idea of a woman in the cockpit scared them. No woman could possibly handle dangerous machinery or function calmly in an emergency. Could that really have happened only twenty years ago? That wasn't even during the dark ages before the women's movement, but during a time of society's supposed awakening.

Snowflakes are gliding like a thousand skaters on a pond, in a Russian novel set in a previous era, when women wore muffs and didn't twirl too fast. Ice is slippery because it melts under the pressure of skating, sledding, skiing, or walking, and becomes a watery lubricant for the briefest moment, growing hot and fast and loose when stressed. Afterward, it cools down, pulls itself together, and re-forms its famously rigid face.

Even in snowfall, winter has distinctive aromas that I enjoy, from pine, cedar, yew, damp bark, decaying leaves, wood smoke, and tree resins, to the air itself. Nothing smells as clean as new snow, which washes the air as it falls, grabbing particulate matter and dragging it to the ground. Afterward the air smells fresh as snowfall and slightly effervescent.

50

On cold sober mornings of no deliverance, when not even the local beauty of several crows, like black velvet appliquéd onto the snow, can distract me, I walk out among the low and mortal. I dwell on earth caky after a hard rain, the breathless sump where a body dislocates forever, that will cure all my aims and ills, and for once I am not afraid, this hour having feasted on the meteoric snow, on tires ringing from a bridge the sounds of a jet in flight, on the mosaic whisperings of a fire (now like cavalry galloping, now like blankets being shaken in the wind), on eggnog thick and custardy as a frappé, on the skull-obsessed Aztecs who carved heads out of rock crystal and gold, on a cream-colored mare with a face like a brioche, on the lisping of the wind, on Aristotle's "wit is well-bred insolence," on fingerprints loopy as a weather map, on the gaudy farewell of autumn whose leaves look like coral eyed through a glass-bottom boat, on highways straight as a statute, on starlings creeping along the barn rafters like Christians in ancient Rome, on winter's unexpected bombs of sunlight.

A garden is never finished, much as it may evolve. In the end, we never complete our own growth, we just keep growing, if we are lucky, until we stop. We don't grow continuously or smoothly or even noticeably at times, but stumblingly, glacially, or at a gallop, without meaning to, or after great effort. We grow because life is growth and we love life not only as an idea, but compulsively, anonymously, in every cell and membrane. Our curiosity is a kind of membrane, too, as are love, ambition, belief, and the many tissues of desire, which lead us from one season to the next and define us in the end. We grow.

51

For months snow ruled, four or five feet of it, drifting like egg whites over the garden. But a gentle thaw settled in on Monday, and all week I've heard its sinusy drip through the gutters, watched snow vanishing like cream. The solitary dribble of an icicle became water torture to my chaos-loving ears. This noon the stone bench grew visible again. Thaw is a new philosophy in the garden. Two crows, flapping hard against the wind to land, look like waiters whisking crumbs from the tablecloth.

A squirrel leaping among the arbors tried to shimmy down a wire holding the bird feeder. After twenty attempts, she settled for the crust of bread I tossed her, but she clearly preferred the sweet swinging seeds overhead. It must have vexed her all day because later she returned, sized up the job, scampered along the arbor, and punched the hell out of the wires until seeds jiggled down like pennies from a bank. Then she hurled herself at the feeder and it crashed to the ground, spilling seeds everywhere. *It's yours,* I said laughing to myself. *You've earned it.*

This has been a glorious year in the garden, though a hard year otherwise, filled with many losses. My garden existed as a sanctuary beyond the world's fevers, comforting me as it has gardeners throughout the ages. All burdens can be dropped at the garden gate, beyond which light sings in the trees and the flower beds patiently wait for attention. Each spring, a garden triumphs over adversity and flourishes again, despite change, dormancy, and any number of setbacks. A garden is always a saga of life and death, mystery and marvel, as another Marvell once eloquently said:

Meanwhile the Mind, from pleasure less,
Withdraws into its happiness:
The Mind, that Ocean where each kind
Does straight its own resemblance find;
Yet it creates, transcending these,
Far other Worlds, and other Seas;
Annihilating all that's made
To a green Thought in a green Shade.

—Andrew Marvell, "The Garden"

52

Is it spring yet? Spring travels north at about thirteen miles a day, which is 47.6 feet per minute, or about 1.23 inches per second. That sounds rather fast, and viewable. I start looking for subtle clues and signs in the snowscape. Weeping willow branches have already started to turn yellow, and the tops of distant trees look dusty pink from new buds. A few cardinals have arrived early to claim the best nesting sites before their rivals return, and I swear I heard the steely twang-and-kazoo of a red-winged blackbird.

In the woods, a skunk cabbage pokes a green spear up through a mound of snow. Reaching down, I spread my fingers near it and feel warmth. Skunk cabbage fires up its own furnace in late winter, by converting taproot starches into sugars, which burn fast and hot, sometimes raising the surrounding air temperature by as much as 20 degrees. We think only of animals as being warm-blooded, but skunk cabbage acts like us in this way: in cold weather it generates body heat. Named after its odor, which smells like rotting flesh—a banquet for hungry flies, its main pollinator—skunk cabbage blooms early in the marshes. But I find other wildflowers already blooming, too. My own local spring begins here in the woods, with trout lilies, skunk cabbage, blue violets, mayapples, triangular white trilliums, pink wild geraniums, and tiny white Dutchman's breeches (whose pant legs are stuffed with nectar). A little later, spring will meander toward the garden.

On the other hand, although I've forced some magnolia branches indoors, the tree hasn't swollen into bloom yet. The apple trees haven't begun to blossom, or the redbud to unfurl its heart-shaped leaves and fuchsia flowers. The lilacs aren't oozing scent. The long smooth curving forsythia branches have turned a dull gold, but it will be weeks before their sunny cascades. On one warm day last week, the first rabbits of the season appeared as if yanked out of a magician's hat. What alarm rang in their veins? Why choose that Monday? No sooner did they awaken than they began furiously courting, chasing one another across the yard, doing pogo-stick dances together in the bouncy minuet I've come to associate with randy rabbits. The squirrels are frantically foraging. A dawn chorus has been acquiring a few new members, especially cardinals hoping to get a jump on other nest seekers. By all the signs, spring is still young in the sap.

A deer wanders into the squirrel feeding oasis and begins to eat the fallen corn below the squirrel toys. Where are the other deer, I wonder. They usually appear in groups. Then a surprise: I see short two-inch horns atop his head. He is a young buck whose antlers will soon grow, and, yes, male deer tend to be loners. When

I toss him an apple, he sniffs at its human smell and retreats into the woods but returns soon for the corn.

Now and then, the squirrels tumbling overhead spook him, but he continues to graze on the treasure of corn. A yearling, strong and muscular like the does, he has the same face and eyes ringed in chamois brown. The squirrels also have outlined eyes, as do we (our eyelids). Eyes are important features to spot right away, to read an animal's mood and intentions. While seven gray squirrels busily chase up and down the trees, the deer finally settles down to a comfortable graze.

Although I'm only twenty feet away, indoors, sitting on a couch in the garden room, the buck is alert to me, but he doesn't seem afraid of my watching. Something at a neighbor's house startles him, though, and he slowly returns to the woods, waits and watches, casts a glance back at the house, and returns cautiously to the corn pile again. Another young buck saunters out of the woods and joins him. Their incipient antlers look like shotgun shells. Then Triangle, the big doe, appears. My heart brightens, since I haven't seen her for months. Are they her yearlings?

From time to time, the three deer eye the corn cobs just out of reach above them, but there's plenty of corn on the ground. Skittish, they get frightened of something real or imaginary and trot into the woods, the doe leading the way toward a neighbor's low fence. When she finds a place she can leap, she lines up squarely and hurdles the fence. The bucks pace nervously. One sniffs at the fence, shifts its weight onto its haunches, tenses its muscles, loses nerve and uncoils, paces back and forth, then takes a savage risk (of breaking a delicate leg) and rockets up, flicking his back hooves as he clears the fence. The smaller second buck trots the long way around through the woods. They'll probably meet up in the forested corridor that meanders for a mile or so between suburban yards and leads to Sapsucker Woods. I hope they don't cross any busy intersections. I've often seen them on lawns in adjoining neighborhoods and been amazed by how far they routinely travel. These woods and yards smell familiar to them. As I said earlier, we're squatters on their land, and we shouldn't be surprised if they drop by for dinner.

In late winter, one imagines the gardens beneath the snow. Paradise exists just out of direct sight, in a backward or slantwise or forward glance, around the crease of a calendar page. Thus I remember the paradise of last summer's roses and anticipate the paradise of thousands of daffodils in spring. All gardeners are optimists, our compass points forward. The garden already exists as a future state of grace I need only await. Why didn't I plant any nasturtiums last summer? In *My Garden (Book)*, Jamaica Kincaid says of their narcotic pleasure: "At night, the smell from them—sweet, like something fermenting that when consumed would make you crazy—was delicious." I can't resist that aromatic delirium; in next year's paradise, I plan to inhale great mounds of nasturtiums at night. This year I completely forgot to plant the king-size ornamental kale that creates a purple-and-green archipelago along the driveway in late autumn. Should I install a Zen garden, in which nothing will change for centuries but the surface of the stones, which may gather lichen and moss? Should I build a gazebo, a structure first created in Persia as a platform for moon watching? Which tulips will have returned? Unfortunately, tulips only bloom for a few years, and I never keep track of which ones I planted when, preferring the surprise of returning familiars and new immigrants to my garden. Each mother bulb I planted in the fall is pregnant with daughter bulbs, which will inherit unequal amounts of her energy. After blooming, the mother will die, and the strongest daughters will bloom the following season. Because the energy is divided each year, in time the bulbs will simply exhaust themselves. So I treat tulips as annuals, enjoy them while they last, don't expect commitment, and count myself lucky if they hang around. As a result, I never know what the complexion of the spring garden will look like, except that there will be an exuberance of petals in a landscape where anarchy rules and any state of green decorum I may achieve is temporary, a flash of control in a wilderness of thieves.

I want time to pool, not race, but tonight I'm madly impatient for the growing season to begin, and the garden, which is a different Eden for every gardener, to reinvent itself as a renewable paradise, if not a permanent one. I don't believe in garden gods, but I do believe in the power of invocation to stir the spirit. Garden of growth, garden of green blood, garden where dappled light and water mix in the trees, crow garden, beetle garden, garden of dreams, garden on the oasis of a life-drenched planet, garden where desire finds form, garden of floral architecture and speckled fawns, garden where wonder is incised on a pebble millions of years old, garden visibly and invisibly teeming, garden of beds and seed parlors, garden of dew and overdue, garden where we plight our troth and ply our trade, garden that tilts the mind into the sacred, fleeting garden, memorial garden, garden abuzz and atwitter, garden where toxins and tonics both thrive, pool garden, cloud garden, garden that's an urn for the soul, garden of roll calls and lists where life tests different recipes, garden where rain falls like manna, garden whose perennial borders are infinite, garden whose customs and taboos make mischief in the mind, garden of snow, mind garden, garden of quartz crystal and siren light.

Any day now spring will revive the blast-frozen valley, tune up the dawn chorus, and hymn the poplars with winter's end. I swear, the willows are yellowing as I look at them. Other trees are glistening with sap, blurring with buds. Squirrels are fidgeting in last year's leavings. Moss has laid down a welcome mat, and red-capped fungi are mustering like British soldiers in a rum confusion of sun and ice. Spring is unlatching its heavy doors, rousting old dusty hibernators from their sleep, and beginning a quiet fumbling with buttons, knots and nubbins, and the bolting ribbons of time, light, and gore. As I walk down to the mailbox, enveloped in mist, birds snitch on twitchy feet in the aspens, morning ghosts between the houses, and the air tastes green at last.

Addendum

What follows is an inventory of the plants in my garden, taken in early September. It doesn't include spring bulbs, because there are so many varieties the list would be endless. I also haven't itemized the 100 or so roses in three rosebeds or the native woodland plants growing in the woodsy walk. However, I thought it might be helpful if I noted the light conditions for each of the following beds, because it's taken a few years to figure out what grows best in what conditions. I'm including it in case you wish to picture what's planted where, or in case you just enjoy the colorful names, as I do.

PLANT INVENTORY, SEPTEMBER 2000

1. Bed Next to Front Walk (Partial Shade)

Rosa x hybrida	Fairy roses: red and pink
Astilbe x arendsii	False spirea
Sedum spectabile	Sedum 'Autumn Joy'
Aquilegia x hybrida	Columbine: self-sown seedlings
Aster novae-angliae	New England aster 'Alma Potschke'
Achillea millefolium	Yarrow 'Cerise Queen'
Eupatorium rugosum	Boneset 'Chocolate'
Lobelia cardinalis	Cardinal flower
Baptisia australis	Blue false indigo
Verbascum chaixii	Chaix mullein
Ligularia przewalskii	The Rocket

2. Raised Bed Next to Driveway (Full Sun)

Rosa x hybrida	Fairy roses
Phlox subulata	Creeping phlox

Iris ensata	Japanese iris 'Batik'
Iris pumila	Dwarf iris 'Cherry Garden'
Aurinia saxatilis	Basket of gold
Aster novi-belgii	Aster 'Patricia Ballard'
Sedum kamtschaticum variegatum	Variegated sedum
Aster laterifolius	Aster 'Prince'
Papaver orientale	Oriental poppy 'Beauty of Livermore,' 'Royal Wedding'
Hypericum calycinum	St. John's wort
Dianthus	Pinks 'Ideal Deep Violet'
Euphorbia polychroma	Cushion spurge
Sedum spectabile	Sedum 'Matrona'
Platycodon grandiflorus	Balloon flower 'Shell Pink'
Chrysanthemum x superbum	Shasta daisy
Rudbeckia fulgida	Black-eyed Susan
Perovskia atriplicifolia	Russian sage
Lavendula angustifolia	Lavender
Iberis sempervirens	Candytuft
Aster novae-angliae	New England aster 'Alma Potschke'
Sedum spurium	Two-row stonecrop
Coreopsis lanceolata	Lanceleaf coreopsis

3. Bed at End of Driveway (Full Sun)

Hemerocallis x hybrida	Daylily 'Stella de Oro'
Rudbeckia fulgida	Black-eyed Susan
Lychnis coronaria	Rose campion
Phlox subulata	Creeping phlox
Aster novi-belgii	Aster 'Professor Kippenburg'
Echinacea purpurea	Purple coneflower
Sedum spectabile	Sedum 'Autumn Joy'
Scabiosa caucasica	Pincushion flower 'QIS Scarlet'
Dendranthema	Chrysanthemum 'Mary Stoker'
Chrysanthemum x superbum	Shasta daisy
Aster dumosus	Aster 'Nesthaekchen'
Aster novae-angliae	New England aster 'Alma Potschke'
Cheiranthus allionii	Wallflower
Knautia macedonica	Knautia
Boltonia asteroides	Thousand-flowered aster
Solidago x hybrida	Goldenrod 'Golden Fleece'

Papaver orientale	Oriental poppy 'Princess Victoria'
Aquilegia x hybrida	Columbine
Buddleia davidii	Butterfly bush 'Black Knight'
Oenothera fruticosa	Sundrops

4. Small Circle Near Driveway Bed (Full Sun)

Achillea millefolium	Yarrow 'Walter Funcke,' 'Fireland,' 'Snowsport'
Hemerocallis x hybrida	Daylilies

5. Large Bed That Wraps Around Front and Side of House (Part Shade, Full Shade)

SHRUBS

Syringa patula	Lilac 'Miss Kim'
Kalmia latifolia	Mountain laurel
Cotinus coggygria	Smoke bush
Pieris japonica	Andromeda
Buddleia davidii	Butterfly bush 'Black Knight,' 'Sungold'
Prunus cistena	Sandcherry
Spirea nipponica	Snowmound spirea
Calycanthus floridus	Carolina allspice
Fothergilla gardenii	Fothergilla
Itea virginica	Virginia willow
Euonymous alatus	Winged wahoo
Forsythia intermedia	Forsythia
Campsis radicans	Trumpet vine
Rosa x hybrida	Roses 'Carefree Delight,' 'All That Jazz,' 'Guy de Maupassant,' 'Blaze'

PERENNIALS

Hemerocallis x hybrida	Daylilies 'Siloam Double Classic,' 'Breathless Beauty,' 'Pandora's Box,' 'Mary Todd,' 'Brocaded Gown,' 'Happy Returns,' 'Stella de Oro'
Lilium orientale	Oriental lilies
Phlox paniculata	Garden phlox
Digitalis purpurea	Foxglove

Eupatorium rugosum	Boneset 'Chocolate'
Aster novae-angliae	New England aster 'Alma Potschke,' 'Honeysong Pink'
Aster novae-belgii	New York aster 'Patricia Ballard'
Rudbeckia fulgida	Black-eyed Susan
Sedum spectabile	Sedum 'Autumn Joy'
Platycodon grandiflorus	Balloon flower
Aquilegia x hybrida	Columbine
Lychnis coronaria	Rose campion
Papaver orientale	Oriental poppy
Perovskia atriplicifolia	Russian sage
Verbascum chaixii	Chaix mullein
Verbascum densiflorum	Mullein
Aster laevis	Aster 'Bluebird'
Polemonium caeruleum	Jacob's ladder
Liatris spicata	Gayfeather 'Floristan White'
Geranium dalmaticum	Hardy geranium
Penstemon barbatus	Penstemon 'Jingle Bells'
Aster x hybrida	Aster 'Astee Roset'
Veronicastrum virginicum	White culver's root
Belamcanda chinensis	Leopard flower
Centaurea macrocephala	Globe centaurea
Artemisia ludoviciana	Wormwood 'Valerie Finnis'
Delphinium elatum	Delphinium 'Astolat,' 'Bluebird,' 'Black Knight'
Sedum telephium	Sedum 'Matrona'
Aquilegia x hybrida	Columbine 'Black Barlow,' 'Blue Barlow'
Veronica spicata	Blue speedwell
Baptisia australis	Blue false indigo
Coreopsis lanceolata	Lanceleaf coreopsis
Chrysanthemum x superbum	Shasta daisy
Veronica spicata	Speedwell 'Icicle'
Dianthus barbatus	Sweet William
Salvia x superba	Salvia 'May Night'
Sidalcea malviflora	Prairie mallow 'Party Girl'
Sedum spectabile	Sedum 'Carmen'
Penstemon x hybrida	Beardtongue 'Husker's Red'
Tiarella wherryi	Foamflower
Astilbe x arendsii	False spirea 'Catherine Deneuve'

Lamiastrum galeobdelon	Archangel 'Herman's Pride'
Lilium orientale	Oriental lily 'Hit Parade,' 'Mediterranee,' 'Begamo,' 'Milano,' 'On Fire'
Phlox stolonifera	Tufted phlox 'Bruce's White'
Phlox divaricata	Wild sweet William 'Fuller's White'
Ligularia przewalskii	The Rocket
Cimicifuga racemosa	Black snakeroot 'Brunette'
Liriope spicata	Variegated lily turf
Digitalis	Foxglove 'Grime's Dwarf'
Lobelia siphilitica	Great blue lobelia
Silene maritima alba	Silene 'Robin White Breast'
Lobelia gerardii x vedrariensis	Violet lobelia
Stachys officinalis	Betony
Sempervivum tectorum	Hens and chicks
Opuntia humifusa	Hardy cactus 'Rafinesquei'
Thalictrum delavayi	Meadow rue
Dendranthema	Chrysanthemum 'Clara Curtis'
Oenothera fructicosa	Sundrops

ANNUALS

Lopezia racemosa	Dragonfly plant
Zinnia elegans	Zinnia 'Dreamland Mix'
Antirrhinum majus	Snapdragon
Angelonia angustifolia	Angel mist
Amaranthus caudatus	Love-lies-bleeding
Impatiens x hybrida	New Guinea impatiens
Clerodendron	Glory bower
Lobelia	Lobelia 'Crystal Palace'
Mimulus	Monkey flower
Viola	Violet 'Bowles Black'
Nigella	Love-in-a-Mist
Cosmos sulphureus	Orange cosmos
Tagetes erecta	African marigold

6. Garden Under Apple Trees (Partial to Full Shade)

SHRUBS

Rhododendron catawbiense	Rhododendron 'Boursalt,' 'Olga Mezzit'
Rhododendron roseum elegans	Rhododendron
Ilex aquifolium	English holly
Leucothoe fontanesiana	Drooping leucothoe
Pieris japonica	Andromeda

PERENNIALS

Polemonium Brise d'Anjou	Variegated Jacob's ladder
Primula japonica	Japanese primrose

7. Shade Garden Under Apple Trees (Full Shade)

Lobelia siphilitica	Blue lobelia
Lobelia cardinalis	Cardinal flower
Eupatorium rugosum	Joe-Pye weed 'Chocolate'
Ligularia stenocephala	Big Leaf Golden Ray
Lamiastrum goleobdolon	Archangel
Ajuga reptans	Bugle
Clematis integrifolia	Clematis
Tradescantia x andersoniana	Dayflower
Hosta x hybrida	Hosta 'Royal Standard, ' 'Guacamole,' 'Blue Moon,' 'Cadet,' 'Fringe Benefit,' 'Regal Splendor,' 'Golden Tiara,' 'France Williams,' 'Gold Standard,' 'Invincible'
Hosta sieboldiana elegans	Plantain lily
Physostegia virginiana	Obedient plant 'Variegata'
Viola labradorica	Labrador violet
Geranium maculatum	Wild geranium
Myosotis biennis	Forget-me-not
Heuchera sanguinea	Coral bells 'Palace Purple'
Astilbe chinensis	Chinese astilbe 'Pumila'
Astilbe x arendsii	Astilbe
Sedum spectabile	Sedum 'Autumn Joy,' 'Carmen'
Helenium autumnale	Sneezeweed

Geranium endressii	Pyrenean cranesbill
Lilium orientale	Oriental lily 'Stargazer,' 'Begamo,' 'Hit Parade'
Aquilegia x hybrida	Columbine
Ligularia przewalskii	The Rocket
Matteuccia struthiopteris	Ostrich fern
Digitalis purpurea	Foxglove
Hemerocallis x hybrida	Daylily 'Stella de Oro,' 'Chicago Apache,' 'Nob Hill,' 'Bittersweet Honey,' 'September Gold'
Carlina vulgaris	Carlina 'Silver Star'
Aster novae-angliae	Aster 'Harrington's Pink'
Helianthus salicifolius	Willowleaf sunflower
Clematis	Clematis 'Madame Julie Correvon,' 'John Warren'
Mentha officinalis	Curly mint 'Crispa'
Oenothera fruticosa	Sundrops
Lysimachia mummularia	Creeping Jenny
Vinca minor	Myrtle
Onoclea sensibilis	Sensitive fern

8. Poolside Garden (Full Sun)

Lavandula angustifolia	Lavender
Sedum spectabile	Sedum 'Frosty Morn'
Hemerocallis x hybrida	Daylily
Hibiscus	Hibiscus 'Southern Belle'
Lamium maculatum	Spotted nettle
Caryopteris x clandonensis	Blue mist spirea
Lilium orientale	Oriental lily
Phlox paniculata	Garden phlox (white)
Sedum spectabile	Sedum 'Autumn Joy'
Phlox subulata	Creeping phlox
Echinacea purpurea	Purple coneflower

9. Zen Garden (Sun, Dry Soil)

Lilium orientale	Oriental lilies
Lilium asiaticum	Asiatic lilies
Lilium tigrinum	Tiger lilies

Hosta	Hosta
Sedum spectabile	Sedum 'Autumn Joy,' 'Carmen,' 'Matrona'
Anemone x hybrida	Japanese anemone

10. Long Border at Foot of Yard (Partial Shade)

Hemerocallis x hybrida	Daylilies
Lythrum salicaria	Purple loosestrife
Ligularia przewalskii	The Rocket
Ligularia stenocephala	Big Leaf Golden Ray
Phlox paniculata	Garden phlox (white, pink, lavender)
Helenium autumnale	Sneezeweed
Lobelia siphilitica	Blue lobelia
Aster species	Asters, wild and cultivated
Paeonia suffruticosa	Peony
Lobelia x hybrida	Purple lobelia
Eupatorium purpureum	Joe-Pye weed
Eupatorium cannibium	Hemp agrimony
Oenothera fruticosa	Sundrops
Astilbe x arendsii	Astilbe
Chrysanthemum superbum	Shasta daisy
Chelone glabra	Turtlehead
Onoclea sensibilis	Sensitive fern
Solidago species	Goldenrod

11. New Bed (Not Fully Planted)

SHRUBS AND TREES

Malus	Crab apple 'Radiant'
Salix integra	Variegated Japanese willow
Salix discolor rosea	French pink pussy willow
Hibiscus syriacus	Rose of Sharon
Euonymus alata	Winged euonymus
Forsythia x intermedia	Forsythia
Rosa x hybrida	Rose 'Simplicity'
Hosta	Hosta

12. Garden Next to Patio (Full Sun)

Hosta	Hosta
Hemerocallis	Daylilies
Lobelia siphilitica	Blue lobelia
Rudbeckia fulgida	Black-eyed Susan
Caltha palustris	Marsh marigold
Dendranthema	Chrysanthemum 'Clara Curtis'
Yucca filamentosa	Variegated yucca
Aster species	Aster 'Professor Kippenburg,' 'Woods Light Blue,' 'Alma Potschke'
Aster lateriflorus	Aster 'Lady in Black'
Ligularia przewalskii	The Rocket
Ligularia dentata	Big Leaf Golden Ray
Achillea millefolium	Yarrow 'Paprika'
Hyssopus officinalis	Hyssop
Anthemis kelwayi	Golden marguerite
Calamintha nepetoides	Calamint
Helianthus helianthoides	Perennial sunflower 'Flora Pleno'
Sedum spectabile	Sedum 'Carmen'
Sedum sieboldii	Sedum 'Vera Jameson'

13. Squirrel Garden (Partial to Full Shade)

Hemerocallis hybrids	Daylilies 'Manila Moon,' 'September Gold,' 'Pandora's Box,' 'Happy Returns,' 'Mary Todd'
Eupatorium rugosum	Joe-Pye weed 'Chocolate'
Ligularia przewalskii	The Rocket
Ligularia dentata	Big Leaf Golden Ray
Sedum spectabile	Sedum 'Autumn Joy'
Rudbeckia fulgida	Black-eyed Susan
Eupatorium purpureum	Joe-Pye weed
Hakonechloa macra	Japanese variegated grass
Lobelia cardinalis	Cardinal flower
Sanguisorba canadensis	Great burnet
Primula denticulata	Primrose
Primula japonica	Japanese primrose
Astilbe chinensis	Chinese astilbe 'Pumila'
Dicentra spectabilis	Bleeding heart

Monarda didyma	Bee balm
Aster novae-angliae	New England aster 'Alma Potschke'
Lobelia siphilitica	Blue lobelia
Lobelia hybrid	Purple lobelia
Polemonium caerulum	Variegated Jacob's ladder 'Brise d'Anjou'
Pieris japonica	Andromeda 'Mountain Fire' (shrub)

14. Small Garden Opposite Squirrel Garden (Shade, Dry Soil)

Athyrium nipponicum	Japanese painted fern
Hakonechloa macra	Japanese variegated grass
Sedum spurium	Stonecrop 'Tricolor'
Myosotis biennis	Forget-me-not
Primula polyanthus	Primrose
Brunnera macrophylla	Bugloss
Galium odoratum	Sweet woodruff

15. The Glade (Heavy Shade)

Hemerocallis hybrids	Daylilies
Hosta hybrids	Hosta
Sedum spectabile	Stonecrop 'Autumn Joy,' 'Rosy Glow'
Primula denticulata	Primrose
Pulmonaria officinalis	Lungwort
Lamium maculatum	Spotted nettle
Physostegia virginiana	Obedient plant
Trollius europaeus	Globeflower

16. Library Garden (Partial Shade)

Hemerocallis	Daylilies
Astilbe arendsii	Astilbe
Astilbe chinensis	Chinese astilbe 'Pumila'
Lobelia cardinalis	Cardinal flower
Thalictrum aquilegifolium	Meadow rue
Rudbeckia fulgida	Black-eyed Susan
Pulmonaria officinalis	Lungwort

Physostegia virginiana	Obedient plant
Chelone oblique	Turtlehead
Trollius europaeus	Globeflower
Polemonium caeruleum	Jacob's ladder
Helleborus foetidus	Lenten rose
Primula japonica	Japanese primrose
Saxifraga	Saxifrage
Tussilago farfara	Coltsfoot
Aquilegia hybrids	Columbine

17. Border Next to Driveway (Full Sun)

Coreopsis verticillata	Threadleaf coreopsis
Paeonia suffruticosa	Yellow peony
Miscanthus zebrinus	Zebra grass
Rudbeckia fulgida	Black-eyed Susan
Hemerocallis hybrid	Daylily 'Stella de Oro'
Heliopsis helianthoides	False sunflower
Erianthus ravennae	Plume grass
Digitalis lutea	Yellow foxglove
Rudbeckia triloba	Yellow coneflower
Coreopsis lanceolata	Lanceleaf coreopsis
Achillea filipendulina	Gold yarrow
Iris siberica	Siberian iris
Miscanthus sinensis	Silver feather grass
Carex	Sedge

Acknowledgments

Many thanks to Carrie Szewczyk for her invaluable assistance, which allowed me to spend my days writing and in the garden while I was recovering from a cycling accident.

Chrys Gardener was kind enough to read the manuscript, looking for infelicities. Both she and Bill Carini are lovely spirits who have helped to bring joy to both the garden and its owner.

A few short passages in this book appeared in a different form in *Civilization,* the *New York Times, Parade,* and *Victoria.* I'm grateful to their editors for their welcome.

Heartfelt thanks to Paul West and Jeanne Mackin, who cast loving eyes over the manuscript; to my agent, Virginia Barber, for her gallantry and good works; and to my editor, Terry Karten, for her diligent and inspired enthusiasm.

Permissions

Grateful acknowledgment is given to the following for their permission to use excerpts:

Page vii: From *Watt* by Samuel Beckett. Copyright © 1953 by Samuel Beckett. Used by permission of Grove/Atlantic, Inc.

Page 11: From *The Poetry of Robert Frost* edited by Edward Connery Lathem, copyright 1936 by Robert Frost, copyright © 1964 by Lesley Frost Ballantine, © 1969 by Henry Holt and Company, LLC. Reprinted by permission of Henry Holt and Company, LLC.

Pages 13–14: From *Portraits of the Rain Forest* by Adrian Forsyth, copyright © 1990. Reprinted by permission of Firefly Books Ltd.

Page 36: From *On Growth and Form* by D'Arcy Wentworth Thompson. Reprinted with the permission of Cambridge University Press.

Page 37: "The Force That Through the Green Fuse Drives the Flower" by Dylan Thomas, from *The Poems of Dylan Thomas*, copyright 1939 by New Directions Publishing Corp. Reprinted by permission of New Directions Publishing Corp.

Page 83: Reprinted with permission from *Biodiversity.* Copyright © 1998 by the National Academy of Science. Courtesy of the National Academy Press, Washington, D.C.

Page 85: From *What Gardens Mean* by Stephanie Ross. Used with the permission of the University of Chicago Press.

Page 88: From *Gardens of Obsession* by Gordon Taylor and Guy Cooper, copyright © 1999, Casswell & Co.

Page 108: "Out on the Lawn I Lie in Bed" from *W. H. Auden: Collected Poems* by W. H. Auden, copyright © 1976 by Edward Mendelson, William Meredith, and Monroe K. Spears, Executors of the Estate of W. H. Auden. Used by permission of Random House, Inc.

About the Author

Poet, essayist, and naturalist, Diane Ackerman was born in Waukegan, Illinois. *Cultivating Delight: A Natural History of My Garden* is her most recent book. She is also the author of many highly acclaimed works of nonfiction including the bestselling *A Natural History of the Senses* as well as the volumes *Deep Play, A Slender Thread, The Rarest of the Rare, A Natural History of Love, The Moon by Whale Light,* and a memoir of flying, *On Extended Wings.*

Her poetry has appeared in literary journals and magazines, and has been collected into several volumes including *Jaguar of Sweet Laughter: New and Selected Poems* and most recently, *I Praise My Destroyer*. She is also writing a series of nature books for children, the first two of which are *Monk Seal Hideaway* and *Bats: Shadows in the Night.* She is co-editor (with Jeanne Mackin) of an anthology, *The Book of Love*.

Diane Ackerman has received many prizes and awards, among them the John Burroughs Nature Award and the Lavan Poetry Prize. She has taught at several universities, including Columbia, New York University, and Cornell, where she is currently Visiting Professor at the Society for the Humanities. A five-hour PBS television series, inspired by *A Natural History of the Senses*, aired in 1995 with Ms. Ackerman as host. She has the unusual distinction of having a molecule named after her (dianeackerone). Diane Ackerman lives in upstate New York.